T0230479

# Computational Modeling of Neural Activities for Statistical Inference

Antonio Kolossa

# Computational Modeling of Neural Activities for Statistical Inference

 Springer

Antonio Kolossa
Institut für Nachrichtentechnik
Technische Universität Braunschweig
Braunschweig
Germany

ISBN 978-3-319-81243-4        ISBN 978-3-319-32285-8    (eBook)
DOI 10.1007/978-3-319-32285-8

Printed on acid-free paper

This Springer imprint is published by Springer Nature
The registered company is Springer International Publishing AG Switzerland

# Foreword

Thomas Bayes (around 1701–1761) took a deep interest in probability theory. His seminal essay on inverse probabilities was published in the *Philosophical Transactions of the Royal Society* of London in 1763, 2 years after Bayes had passed away. What we know today as Bayes' theorem has become fundamental to many scientific disciplines such as engineering, natural sciences, neurosciences, cognitive sciences, statistics, and beyond. The field of decision and estimation theory is totally centered around Bayes' theorem today.

The background of this book is the encounter of two contemporary followers of Bayes: Prof. Dr. rer. soc. Bruno Kopp, affiliated to the Medizinische Hochschule Hannover, Hannover, Germany, and Prof. Dr.-Ing. Tim Fingscheidt, Technische Universität Braunschweig, Braunschweig, Germany. While Kopp's research focuses on understanding the principles of predictive learning in mind and brain, Fingscheidt's research interests cover signal processing and machine learning, mostly with applications in speech. They both believed that event-related potentials (ERPs)—i.e., scalp-recorded real-time proxies of cortical activities—should be predictable by computational methods, and that the pursuit of this modeling effort could create an encompassing theory of how the mind and the brain work. So far the vision. The author of this book, Antonio Kolossa, then sought and found brilliant solutions for mathematically rigorous explanations of how the mind and the brain work—relying on Bayesian methods throughout. Beyond the initial vision of an encompassing theory of mind and brain, Kolossa applied modern Bayesian model comparison techniques to substantiate his discoveries.

These discoveries concern the identification of computational models of cognition from ERPs. Kolossa presents a detailed description of two studies, both of which implied Bayesian model comparison techniques. A model which comprises multiple digital filters yielded the best account for single-trial variability of ERPs that were measured in an oddball task in a first study. A Bayesian belief updating model yielded the best account for single-trial variability of ERPs that were measured in an urn-ball task in a second study. This second study provided evidence for the Bayesian brain hypothesis and for neural probability weighting. The work also lays the methodological ground for the emerging fields of computational cognitive neuroscience, computational psychiatry, and computational neurology.

Tim Fingscheidt
Bruno Kopp

# Preface

The human brain is constantly exposed to variable sensory information about the surrounding environment. How does the brain integrate this information to make reliable inferences and predictions as the basis for decision making? If the available information is to be used in an optimal manner, sophisticated statistical methods need to be employed. The question which methods the brain uses to come to decisions and predictions is still unsolved and one of the most exciting questions of neuroscience today (Bach 2014; Lisman 2015).

In the statistical field, Bayesian decision theory combines all available information in an optimal fashion and thus offers a useful theoretical framework for explaining probabilistic inference by humans (Jaynes 2003; Robert 2007). In this framework prior beliefs are constantly updated to posterior beliefs in light of observed data according to Bayes' theorem (Baldi and Itti 2010). Thus, the Bayesian brain hypothesis, which states that the brain codes and computes Bayesian probabilities, has been proposed and is increasingly recognized as providing a framework for investigating cognitive brain functions (Kersten et al. 2004; Knill and Pouget 2004; Friston 2005; Doya et al. 2007; Gold and Shadlen 2007; Kopp 2008; Friston 2010; Bach and Dolan 2012).

Predictive coding theories of cortical functions and the free energy principle instantiate the Bayesian brain hypothesis (Friston 2002, 2010). They are widely applied frameworks for functional neuroimaging and electrophysiological studies of sensory cortical processing (Summerfield et al. 2006; Garrido et al. 2009; Summerfield and Egner 2009; Egner et al. 2010; Rauss et al. 2011; Winkler and Czigler 2011; Lieder et al. 2013). Put simply, these theories state that the brain tries to minimize any "surprise" or prediction error about sensory input. Specifically, predictive coding theories propose that the brain maintains an internal model of the world which it updates in dependence on the surprise about a current stimulus, in order to minimize the surprise about future stimuli (Friston 2002; Friston 2005; Spratling 2010). While earlier research provided results that are consistent with the Bayesian brain hypothesis (Hampton et al. 2006; Ostwald et al. 2012; Vilares et al. 2012; Lieder et al. 2013), definite empirical support is surprisingly scarce, and no unchallenged conclusion about the utility of the Bayesian brain hypothesis as a theoretical framework for explaining cognitive functions of the brain has been achieved so far (Clark 2013).

This work aims at filling this gap by collecting key experimental data in support of the Bayesian brain hypothesis. Which means are necessary to achieve this goal? First, some window into the brain is needed. The electroencephalogram (EEG), which is the signal of the electrical fields of the brain, was first presented nearly 90 years ago with the potential of providing this window (Berger 1929). Although the EEG data are severely corrupted by noise (Schimmel 1967), they provide signals of neural activity with a high temporal resolution (Makeig et al. 2004), which enables studies of brain signals in direct relation with complex cognitive tasks (da Silva 2013) and makes the EEG a useful tool for brain imaging (Michel and Murray 2012). From all the activity which can be seen in the brain, the so-called event-related potentials (ERPs) are particularly useful for a better understanding of brain functions. They are the "scalp-recorded neural activity that is generated in a given neuroanatomical module when a specific computation operation is performed." (Luck 2014). This implies that by understanding the amplitude fluctuations of these ERPs, the computations of the brain themselves can be deduced.

Thus, the goal of this work is to better understand the probabilistic reasoning of humans by developing observer models to predict event-related potentials, select the models which best explains the EEG data using Bayesian model selection, and make deductions from the properties of the winning models. Note that this is a meta-Bayesian model-based analysis in the sense that Bayesian model selection is used to choose between Bayes-optimal observer models (Daunizeau et al. 2010; Lieder et al. 2013). Inference about the algorithms employed by the brain is then based on the winning models (Mars et al. 2012). A framework of Bayesian updating and predictive surprise is used for dissociating the functions underlying ERP amplitude fluctuations. Bayesian updating refers to changes in probability distributions given new observations and can be differentiated into Bayesian surprise, which constitutes the changes in beliefs about hidden states, and postdictive surprise, which represents the changes in predictions over observable events. In contrast, predictive surprise equals the surprise about observations under their current probabilities.

In a first step, the P300 component of event-related brain potentials is investigated. The P300 is a *positive* potential that is typically measured at parietal scalp regions in a time interval starting 300 ms after an unforeseeable stimulus is presented, and has long been in the focus of research concerned with the brain's ability to infer statistical regularities of the environment (Sutton et al. 1965). It reflects the degree of surprise related to the processing of sensory input in a way that "surprising events elicit a large P300 component." (Donchin 1981). A variant of the well-established oddball task (Ritter and Vaughan 1969) is used to collect the EEG data for developing and testing a digital filtering model (DIF), which fuses properties of the most popular and comprehensive model of P300 amplitude fluctuations with a completely computational model. These were proposed by Squires et al. (1976) and Mars et al. (2008), respectively. While Squires et al.'s model remains descriptive,

Mars et al.'s observer simply integrates the sensory data over an infinitely long period of time and cannot explain the well-documented effects of the recent stimulus sequence on the P300 (Squires et al. 1976; Leuthold and Sommer 1993). The model selection results show that P300 amplitude fluctuations are best explained by predictive surprise based on the DIF model, which provides direct evidence for the coding of probability distributions in the human brain. Evidence for the updating of the probability distributions is, however, only implicit.

In the next step, the analyses are extended to enclose a total of four temporally and regionally distinguishable ERPs in order to find neural traces for the actual updating of the probability distributions as well as their presence. These ERPs are the frontocentrally distributed N250 (Hillyard and Picton 1987), the anteriorly distributed P3a, the parietally distributed P3b, and the posteriorly distributed slow wave (SW) (Matsuda and Nittono 2015). The P3a and P3b are dissociable components of the P300 (Polich 2007), while the P3a, P3b, and SW make up the so-called late positive complex (Sutton and Ruchkin 1984; Dien et al. 2004). A variant of the urn-ball task (Phillips and Edwards 1966) is introduced, which was specifically designed to represent Bayes' theorem. A Bayesian observer model is then proposed from which a belief distribution over hidden states and a prediction distribution over observable events are derived.

Additionally, it is investigated whether observer models that incorporate non-linear probability weighting outperform their versions without weighting when predicting ERP amplitude fluctuations. This nonlinear weighting of probabilities was originally reported by prospect theory, which is a famous theory of economic decision behavior (Kahneman and Tversky 1979; Tversky and Kahneman 1992; Fox and Poldrack 2009). The model selection results show that the ERP components of the late positive complex (P3a, P3b, Slow Wave) provide dissociable measures of Bayesian updating and predictive surprise based on the Bayesian observer, while for the N250 predictive surprise based on the DIF model proved superior. These results indicate that the ERP components reflect distinct neural computations and provide evidence for the coding and computing of probability distributions.

The structure of this work is as follows: Chap. 1 introduces basic principles of ERP research which comprise the data acquisition methods used in this work, signal-to-noise ratio estimation for event-related potentials, and the important concept of circularity in data analyses. It further details the framework of Bayesian updating and predictive surprise and the concept of probability weighting functions. Chapter 2 introduces the parametric empirical Bayes methods and variational free energy used for model estimation and selection. It first motivates the use of these methods before thoroughly detailing their computation. In addition, an example experiment analyzes the performance of these methods for single subjects and group studies in light of the signal-to-noise ratio of the data.

Chapter 3 introduces the oddball task and the models taken from the literature before giving a detailed derivation of the digital filtering (DIF) model. The results are displayed and discussed for conventional ERP analyses as well as model-based trial-by-trial analyses. Next, Chap. 4 introduces the urn-ball task and the Bayesian observer model. Probability weighting functions are applied to the Bayesian observer model and to the DIF model. The discussion of the results for conventional ERP analyses and model-based trial-by-trial analyses concludes this chapter. Finally, Chap. 5 summarizes this work, draws the main conclusions, and closes with an outlook.

# Acknowledgment

This work would not have been possible without the support of several people. First, I want to thank Prof. Dr.-Ing. Tim Fingscheidt and Prof. Dr. rer. soc. Bruno Kopp, whose shared vision was the basis of this work.

Professor Fingscheidt gave me the opportunity to work toward the fulfillment of this vision. He taught me the ways of the researcher and pushed me to work to the best of my abilities by constantly challenging my work and giving me constructive feedback. Professor Kopp constantly motivated me to go on and patiently pointed me toward new leads to follow whenever I deemed myself in a dead end. Together, Prof. Fingscheidt and Prof. Kopp provided inspirations based on principles of machine learning as well as insights into the workings of mind and brain. Without their complementary inputs the aim of computational modeling of cognitive brain functions would have been unreachable.

I would further like to thank Prof. Dr. rer. nat. Karl-Joachim Wirths and my aunt Cosima Meyer for proofreading this work and their helpful comments. I am very grateful toward my grandparents Annemarie and Friedrich Meyer, without whom I would not have tread the paths in life that led me here. They constantly supported me and encouraged me to follow my dreams.

Finally, my special thanks go to my lovely wife Christine. Our time together is so precious I think of each moment "Verweile doch! Du bist so schön!"

Braunschweig                                                          Antonio Kolossa
February 2016

# Contents

# Symbols

Not emphasized letters refer to scalars, bold lowercase letters to (column) vectors, bold capital letters to matrices, the superscript $[\ ]^T$ denotes the transpose, $|\cdot|$ the determinant of a matrix, the number of elements in a set, or the absolute value of a scalar, respectively, while $\text{tr}\{\cdot\}$ denotes the trace of a matrix, and $\langle\cdot\rangle$ is the expectation operator. The symbol (●) signifies a frequent event, (●) a rare event, and (●) the average over both types of events.

## Roman Letters

| | |
|---|---|
| $a$ | Analog signal index |
| a | Parameter controlling the slope of the optimal choice function |
| $b$ | Block index |
| b | Parameter controlling the bias of the optimal choice function |
| $c$ | Experimental condition |
| $c(n)$ | Count function of the DIF model |
| $\breve{c}(n)$ | Count function of the SQU model |
| $\tilde{c}(n)$ | Count function of the MAR model |
| $\tilde{c}(n)$ | Count function of the OST model |
| $\tilde{c}'(v)$ | Temporary count function of the OST model |
| $d_k(n)$ | Input time sequence for the SQU and MAR models |
| $d_{\text{sym}}$ | Symmetric distortion |
| e | Electrode index |
| $f$ | Frequency |
| $f_{\text{sp}}$ | Stimulus presentation rate |
| $g_k(n)$ | Input signal of the DIF model |
| $\tilde{g}_k(n)$ | Input time sequence of the OST model |
| $h(n)$ | Discrete time impulse response |
| **h** | Gradient vector |

| $i$ | Level of the GLM |
|---|---|
| $i'$ | Index for weights of the DIF model |
| $i_m(n)$ | Sequence capturing the surprise from model $m$ |
| $i''_m(n)$ | $i_m(n)$ normalized to zero mean and unit variance |
| $k$ | Event |
| $\log(\mathrm{BF})$ | Log-Bayes factor |
| $\log(\mathrm{GBF})$ | Group log-Bayes factor |
| $\log \mathrm{p}(\mathbf{y}|\boldsymbol{\lambda})$ | Log-likelihood of the data conditional on the hyperparameters |
| $\log \mathrm{p}(\mathbf{y}|m)$ | Log-likelihood of the data conditional on model $m$ |
| $n$ | Trial index |
| $n'$ | Sample index |
| $n'_n$ | Sample index of stimulus presentation on trial $n$ |
| $\ell$ | Subject index |
| $m$ | Model index |
| $m'$ | Relative sample index within an epoch |
| $o$ | Observation |
| $\mathbf{o}_1^n$ | Sequence of observations |
| $\mathrm{p}(\chi)$ | Probability density function over $\chi$ |
| $\mathrm{p}(\mathbf{y}|m)$ | Likelihood of the data conditional on model $m$ |
| $\mathrm{p}(\widetilde{\boldsymbol{\theta}}|\mathbf{y})$ | Probability density function of the parameters of the augmented GLM conditional on the data |
| $q$ | Hidden state |
| $r$ | Predictor index in the GLM |
| $s$ | Sampled signal index |
| $s(n)$ | Clean signal |
| $\widehat{s}(n)$ | Estimate of $s(n)$ |
| $t$ | Time index |
| $u$ | State |
| $w(\cdot)$ | Probability weighting function |
| $(\cdot)^{(w)}$ | Abbreviated form of the probability weighting function $w(\cdot)$ |
| $x(n)$ | Predictor |
| $y(n)$ | Measured signal, sequence of ERP amplitudes |
| $\widetilde{\mathbf{y}}$ | Augmented data vector of the collapsed GLM |
| $y_a(t)$ | Analog continuous time signal |
| $y_n(m')$ | Baseline corrected epoch of the sampled signal |
| $\bar{y}_k(m')$ | ERP waveform |
| $y(n')$ | Quantized sampled signal |
| $y'_n(m')$ | Epoch of the sampled signal |
| $y''(n)$ | $y(n)$ normalized to zero mean and unit variance |
| $A$ | Index denoting alternation expectation |
| $B$ | Number of blocks |
| $\mathcal{B}$ | Set of blocks |
| $C$ | Normalizing constant |
| $C_{\mathrm{L},n}$ | Dynamic normalizing value |

| $\mathcal{C}$ | Set of experimental conditions |
| $\mathrm{D_{KL}}$ | Kullback–Leibler divergence |
| $\mathcal{E}$ | Set of electrodes |
| $\mathrm{E}_k$ | Expectancy for event $k$ |
| $H(f)$ | Transfer function of a digital filter |
| $F$ | Variational free energy |
| $F_{\tilde{\theta}}$ | Free energy |
| $\mathbf{G}$ | Matrix used for abbreviated notation |
| $\mathbf{I}$ | Identity matrix |
| $I_B$ | Bayesian updating; Bayesian surprise; postdictive surprise |
| $I_H$ | Entropy; average predictive surprise |
| $I_P$ | Predictive surprise |
| $\mathbf{H}$ | Fisher's information matrix |
| $H_0$ | Null hypothesis |
| $K$ | Number of types of events |
| $\mathcal{K}$ | Set of types of events |
| $L$ | Number of subjects |
| $\mathrm{L}$ | Index denoting long-term memory |
| $\mathcal{L}_{k|u}$ | Likelihood for event $k$ given state $u$ |
| $\mathrm{Lo}(\cdot)$ | Log-odds function |
| $M$ | Number of models in the model space |
| $\mathcal{M}$ | Model space |
| $N$ | Number of trials |
| $N_{\mathrm{alt}}$ | Number of consecutive stimulus alternations |
| $N_{\mathrm{depth}}$ | Memory span |
| $\mathcal{N}(\mu, \sigma)$ | Normal distribution with mean $\mu$ and standard deviation $\sigma$ |
| $\mathrm{P}_0$ | Crossover point of the probability weighting function |
| $\mathrm{P}_k$ | Marginal probability for event $k$ |
| $\mathrm{P}_k(n)$ | Probability for observation $o$ being $k$ |
| $\mathrm{P}_k(n+1)$ | Prediction for observation $o$ being $k$ |
| $\mathrm{P}_u$ | Initial prior probability for state $q$ being $u$ |
| $\mathrm{P}_u(n-1)$ | Prior belief for the hidden state $q$ being $u$ |
| $\mathrm{P}_u(n)$ | Posterior belief for the hidden state $q$ being $u$ |
| $\overline{\mathrm{P}}_u(n)$ | Average posterior probability |
| $\mathrm{P}(m)$ | Prior probability of model $m$ |
| $\mathrm{P}(m|\mathbf{y})$ | Posterior model probability of model $m$ |
| $P_s$ | Power of the clean signal |
| $\widehat{P}_s$ | Estimate of $P_s$ |
| $P_\epsilon$ | Power of the noise |
| $\widehat{P}_\epsilon$ | Estimate of $P_\epsilon$ |
| $P_y$ | Power of the noisy signal |
| $\mathbf{Q}_i$ | Augmented identity matrix |
| $R$ | Number of different predictors in the GLM |
| $\mathcal{R}$ | Set of different predictors |

| S | Index denoting short-term memory |
| SNR | Signal-to-noise ratio of the ERP amplitudes |
| $\widehat{\text{SNR}}$ | Estimate of the signal-to-noise ratio of the ERP amplitudes |
| $T_s$ | Sampling period |
| $\mathcal{T}_{\text{ERP}}$ | ERP-specific time interval |
| $U$ | Number of types of states |
| $\mathcal{U}$ | Set of types of states |
| $\mathbf{X}$ | Design matrix |
| $\widetilde{\mathbf{X}}$ | Augmented design matrix of the collapsed GLM |

# Greek Letters

| $\alpha$ | Memory weighting parameter of the DIF model |
| $\beta$ | Time constant controlling memory length |
| $\beta_{\text{L},n}$ | Time-dependent value controlling memory length |
| $\gamma$ | Constant controlling exponential forgetting |
| $\gamma_{\text{A}}$ | FIR filter coefficient |
| $\gamma_{\text{L},n}$ | Time-dependent factor controlling exponential forgetting |
| $\gamma'$ | Effective filter coefficient |
| $\delta$ | Choice |
| $\epsilon$ | Error and noise |
| $\varepsilon$ | Step function |
| $\widetilde{\epsilon}$ | Augmented error vector of the collapsed GLM |
| $\zeta$ | Shape parameter of the probability weighting function |
| $\theta$ | Parameter of the GLM |
| $\widetilde{\theta}$ | Augmented parameter vector of the collapsed GLM |
| $\lambda$ | Hyperparameter controlling the error covariances |
| $\mu_{\widetilde{\theta}|y}$ | Vector of the parameter means of the augmented GLM conditional on the data |
| $\nu$ | Sample index |
| $\sigma$ | Standard deviation |
| $\sigma^2$ | Variance |
| $\varsigma_k(n)$ | Sign of the alternation expectancy term of the SQU model |
| $\tau_1, \tau_2$ | Time constants controlling $\beta_{\text{L},n}$ |
| $\tau'$ | Summation index |
| $\upsilon$ | Summation index |
| $\chi$ | Random variable capturing the observation probability |
| $\Gamma$ | Gamma function |
| $\Delta\lambda$ | Update-width of the hyperparameters during EM |
| $\Sigma_\epsilon$ | Error covariance matrix |

| | |
|---|---|
| $\Sigma_{\tilde{\epsilon}}$ | Augmented error covariance matrix of the collapsed GLM |
| $\Sigma_{\tilde{\theta}}$ | Parameter covariance matrix of the augmented GLM |
| $\Sigma_{\tilde{\theta}|y}$ | Parameter covariance matrix of the augmented GLM conditional on the data |
| $\Omega$ | Electrical resistance |

# Abbreviations

| | |
|---|---|
| BEL | Belief distribution |
| BF | Bayes factor |
| BMS | Bayesian model selection |
| BRU | Bayesian reasoning unit |
| DIF | Digital filtering model |
| EEG | Electroencephalogram |
| EM | Expectation maximization |
| ENP | Encompassing model |
| EOG | Electrooculogram |
| ERP | Event-related potential |
| FIR | Finite impulse response |
| GBF | Group Bayes factor |
| GLM | General linear model |
| IIR | Infinite impulse response |
| LQO | Listening quality objective |
| LQS | Listening quality subjective |
| MAR | Model proposed by Mars et al. (2008) |
| MSE | Mean squared error |
| MOS | Mean opinion score |
| NUL | Null model |
| OST | Model proposed by Ostwald et al. (2012) |
| PEB | Parametric empirical Bayes |
| PESQ | Perceptual evaluation of speech quality |
| PMP | Posterior model probability |
| PR | Probabilistic reasoning |
| PRE | Prediction distribution |
| ROI | Region of interest |
| SNR | Signal-to-noise ratio |
| SPM | Statistical parametric mapping |
| SQD | Squared distortion model |
| SQU | Model proposed by Squires et al. (1976) |
| TRU | Ground truth model |
| vWM | Visual working memory |

# Abstract

The Bayesian brain hypothesis is increasingly recognized as offering a theoretical framework for the integration of information by the human brain. This hypothesis simply states that the brain integrates sensory input in a Bayes-optimal manner upon which it bases decisions and predictions. Despite the popularity and importance of this hypothesis for cognitive neuroscience, there is scarce empirical evidence for the underlying neural processes.

This work supplies empirical evidence for the Bayesian brain hypothesis by modeling event-related potentials (ERP) of the human electroencephalogram (EEG) during successive trials in cognitive tasks. The employed observer models compute probability distributions over observable events and hidden states, depending on which are present in the respective tasks. A framework of Bayesian updating and predictive surprise relates the coding and computing of these probability distributions to neural activity. Bayesian updating quantifies the magnitude of change in probability distributions given new observations, while predictive surprise reflects the surprise about new observations given their current probabilities. Bayesian model selection is used to choose the model which best explains the ERP amplitude fluctuations. Decisive evidence for the Bayesian brain hypothesis is collected in two steps:

First, P300 amplitudes are obtained from the EEG signals measured in a variant of the oddball task, which consists solely of observable events. The P300 is an ERP which is typically associated with stimulus probabilities. A new digital filtering model (DIF) is derived inspired by two of the most renowned models of the P300, and is successfully tested against these and a recently published model. The DIF model calculates the probability distributions over observable events and consists of three additive digital filtering processes, representing short-term, long-term, and working memory.

Second, an urn-ball task is introduced, which is specifically designed to represent Bayes' theorem and which consists of observable events and hidden states. A Bayesian observer model is then established from which a prediction distribution over observable events and a belief distribution over hidden states are derived. The ERPs N250, P3a, P3b, and SW are obtained from the EEG signals measured during the performance of the urn-ball task. It will be shown that these ERPs provide dissociable measures of Bayesian updating and predictive surprise, thus indicating that the three components of the late positive complex (P3a, P3b, SW) and the N250 reflect distinct neural computations. While the late positive complex seems to code and compute both probability distributions within the Bayesian observer model, the N250 is best fit by the DIF model, suggesting a distinction between fast memory-based and slow model-based forms of surprise.

The Bayesian observer seems to employ nonlinear weighting of probabilities, which may be one of the reasons why empirical support for the Bayesian brain hypothesis has been so difficult to obtain in previous studies. In its entirety, this work constitutes a decisive step toward a better understanding of the neural coding and computing of probabilities following Bayesian rules.

# Chapter 1
# Basic Principles of ERP Research, Surprise, and Probability Estimation

This section introduces the fundamentals of this work. It starts with a description of the hardware and software which was used to conduct the experiments and further analyses. Next, a method for signal-to-noise ratio estimation of event-related potentials is described, followed by the important concept of circularity in data analyses. It is then shown how evidence for the coding of probability distributions in the brain can be obtained, using a framework that relates random variables to neural activities. Last, an overview on probability weighting by humans is given, the role of which in probabilistic reasoning is investigated in this work. Parts of this chapter have been adapted and extended from Kolossa et al. (2013) and Kolossa et al. (2015).

## 1.1 Data Acquisition and Initial Analysis

This section describes the employed hardware and data acquisition tools used in this work. They were identical for the studies in Chaps. 3 and 4, respectively.

**Stimulus Presentation**

The studies were executed using the Presentation® software (Neurobehavioral Systems, Albany, CA), and an Eizo FlexScan T766 19" (Hakusan, Ishikawa, Japan) computer screen with a refresh rate of $100\,Hz$ and a resolution of $1280 \times 1024$ pixels, on which the stimuli were presented to the subjects on a trial-by-trial basis. The stimuli were $K$ different types of events $k \in \mathcal{K}$ with $\mathcal{K} = \{1, \ldots, K\}$ and formed so-called observations $o(n) = k$ on trials $n \in \{1, \ldots, N\}$, with $N$ as total number of trials.

**Data Acquisition**

The signal $y_{a,e}(t)$ of the continuous electroencephalogram (EEG), with index $a$ denoting the analog signal and index $e \in \{$F7, F3, Fz, F4, F8, FCz, T7, C3, Cz, C4, T8, TP9, P7, P3, Pz, P4, P8, TP10, O1, O2, M1, M2$\}$ the electrode, was sampled with a sampling frequency of $f_s = 250\,Hz$ (i.e., sampling period $T_s = 4\,ms$) and

© Springer International Publishing Switzerland 2016
A. Kolossa, *Computational Modeling of Neural Activities for Statistical Inference*, DOI 10.1007/978-3-319-32285-8_1

**Fig. 1.1** Electrode locations on the scalp for the employed `EasyCap` system

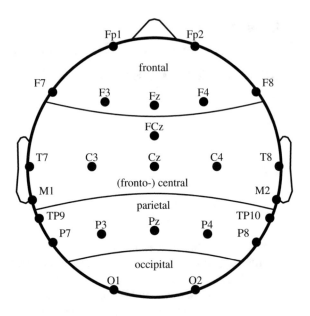

quantized with a 24-bit analog/digital converter, using a `QuickAmps-72` amplifier (Brain Products, Gilching, Germany) and the `Brain Vision Recorder`® Version 1.02 software (Brain Products, Gilching, Germany), yielding the digital signal $y_e(n')$ with sample index $n' \in \mathbb{Z}$. Please note that $y_{a,e}(t = n'T_s) = y_e(n')$ due to the high quantization resolution, and that $y_e(n')$ is commonly used to denote both, the $n'$th sample in the sequence $\{y_e(n')\}$ *and* the whole sequence, depending on the context. An `EasyCap` (EasyCap, Herrsching-Breitbrunn, Germany) was equipped with 24 Ag-AgCl EEG electrodes, whose impedance was kept below $10\,\mathrm{k\Omega}$. Figure 1.1 shows the electrode locations for the frontal (F7, F3, Fz, F4, F8), fronto-central (FCz), central (T7, C3, Cz, C4, T8), parietal (TP9, P7, P3, Pz, P4, P8, TP10), occipital (O1, O2), and mastoid (M1, M2) sites. All EEG electrodes were referenced to average reference during the recording. Subjects were informed about the problem of non-cerebral artifacts, and they were encouraged to reduce the occurrence of movement artifacts (Picton et al. 2000). Ocular artifacts were monitored by means of bipolar pairs of electrodes positioned at the sub- and supraorbital ridges (vertical electrooculogram, vEOG) and at the external ocular canthi (horizontal electrooculogram, hEOG). The EEG and EOG channels were subject to a bandpass of $0.01$–$30\,\mathrm{Hz}$, both being sampled at $f_s = 250\,\mathrm{Hz}$.

**Initial Data Analysis**

Initial offline analysis of the EEG data was performed by means of the `Brain Vision Analyzer`® Version 2.0.1 software (Brain Products, Gilching, Germany). The EEG electrodes were offline re-referenced to average mastoid reference (Luck 2014). Careful manual artifact rejection was performed to discard trials during which eye movements or any other non-cerebral artifacts except blinks had occurred.

Trials in which the subjects selected the wrong behavioral response were excluded from further analyses as well. Deflections in the averaged EOG waveforms were small, indicating that fixation was well maintained for those trials that survived the manual artifact rejection process. Semi-automatic blink detection and the application of an established method for blink artifact removal were employed for blink correction (Gratton et al. 1983). Further data analyses were done using MATLAB 7.11.0 and EEGLAB 11.0.4.3b (Delorme and Makeig 2004), as well as the Statistical Parametric Mapping (SPM8) software (Friston et al. 2007).

**Selection of ERP Data for Further Analyses**

First, a region of interest (ROI) is defined in terms of the electrode of interest which simplifies the following description by dropping the subscript e for the particular electrode. In order to derive the ERP amplitudes $y(n)$ for further analyses, $y(n')$ is first divided into one epoch per trial $n$ yielding

$$\{y'_n(m')\} = \{y(n' = n'_n + m'_{min}), \ldots, y(n' = n'_n + m'_{max})\}, \tag{1.1}$$

with $n'_n$ as sample index, where stimulus $n \in \{1, \ldots, N\}$ was presented and $m' \in [m'_{min}, m'_{max}]$ as relative sample index of the epoch-specific sequence $y'_n(m')$. Notice that $m'_{min}$ is usually a negative value, which captures samples before stimulus presentation, and that $m' = 0$ is the sample index of stimulus presentation within an epoch. Correction of $y'_n(m')$ using the interval $[m'_{base,min}, m'_{base,max}]$ as baseline removes any non-zero bias and yields

$$y_n(m') = y'_n(m') - \frac{1}{||m'_{base,min}| - |m'_{base,max}|| + 1} \sum_{m' = m'_{base,min}}^{m'_{base,max}} y'_n(m'), \tag{1.2}$$

with $|\cdot|$ denoting the absolute value. Next, a region of interest (ROI) is defined in terms of time windows (notice that the time windows have to be transferred to the corresponding sample indices). Application of a search criterion for the ERP component latency $m'_{ERP}$ and additional smoothing of $y_n(m')$ finally yield the ERP amplitudes $y(n)$ (Barceló et al. 2008). Identifying ERP component latencies in single trials is immensely difficult, due to the low signal-to-noise ratio of single-trial EEG data (Blankertz et al. 2002). Thus, in this work a single sample index $m'_{ERP}$ is calculated for all types of events with subsequent averaging, thereby ignoring ERP component latency variability across trials or event types (Luck 2014). Despite this drawback, it is still more reliable than peak detection (Debener et al. 2005) and widely applied in current studies (Mars et al. 2008; Lieder et al. 2013).

An example procedure deriving $y(n)$ based on EEG data collected from subject number 16 in the experiment detailed in Chap. 4 is given in the next paragraph. The actual study-specific values of $m'_{min}$ and $m'_{max}$ as well as the employed search criteria for $m'_{ERP}$ are detailed in Sects. 3.2 and 4.2 for the respective experiments. While the following example contains the sample indices $m'$ as well as the corresponding post-stimulus time $t = m'T_{sa}$, the remaining chapters will only denote time $t$ as this is the meaningful value in ERP research and completely in line with community standards (Luck 2014).

**An Example ERP Amplitude Acquisition Procedure**

This example illustrates the principles underlying the acquisition of ERP amplitudes for the P3b. The EEG of a subject is recorded as described above during the presentation of $N = 200$ stimuli, which are either a frequent ($k = 1$) or a rare ($k = 2$) event. Only electrode e = Pz is considered a part of the ROI and is thus further analyzed. The data of three trials are rejected due to the subject selecting the wrong response, leaving 197 epochs $y'_n(m')$, which are extracted (1.1) in the time window from 100 ms before to 600 ms after stimulus presentation. Note that in order to preserve time linearity between the continuous signal and the signal divided into epochs, this corresponds to $m'_{min} = -24$ and $m'_{max} = 150$, i.e., $\frac{700\,ms}{T_s} = 175$ samples (Delorme and Makeig 2004). The epochs are corrected to obtain $y_n(m')$, using the interval $m' \in [-24, 0]$ as baseline (see (1.2)). The epochs are then separated by their event type, yielding $N_{k=1} = 132$ epochs $y_{n,k=1}(m')$, with $n \in \{1, \ldots, N_{k=1}\}$, during which a frequent event occurred, and $N_{k=2} = 65$ epochs $y_{n,k=2}(m')$, with $n \in \{1, \ldots, N_{k=2}\}$, during which a rare event occurred. Figure 1.2a shows these epochs for the rare events, while Fig. 1.2b shows them for the frequent events. As is typical for EEG data, a visual inspection of Fig. 1.2a versus Fig. 1.2b reveals no modulation of the EEG data by the event type (Luck 2014). In order to make the modulating effect visible and derive the ERP component latency, ERP waveforms $\bar{y}_k(m')$ are created for each event type $k \in \mathcal{K} = \{1, 2\}$ separately as an ensemble average according to

$$\bar{y}_k(m') = \frac{1}{N_k} \sum_{n=1}^{N_k} y_{n,k}(m').\tag{1.3}$$

Figure 1.2c shows these ERP waveforms with the P3b-typical time interval $\mathcal{T}_{P3b}$ from 300 to 400 ms after stimulus presentation highlighted in gray. Visual inspection of Fig. 1.2c reveals an obvious difference between the waveform for the frequent event (——) and the waveform for the rare event (- - -) within the indicated time interval. Notice that larger amplitudes in response to the rare event than in response to the frequent event thus confirming the P3b as a surprise response of the cortex (Donchin 1981). The ERP component latency $m'_{P3b}$, which yields $t_{P3b} = m'_{P3b}T_s$, is calculated as the latency of maximum difference between the two ERP waveforms via

$$m'_{P3b} = \arg\max_{m'} \left\{ \left| \bar{y}_{k=1}(m') - \bar{y}_{k=2}(m') \right| \right\},\tag{1.4}$$

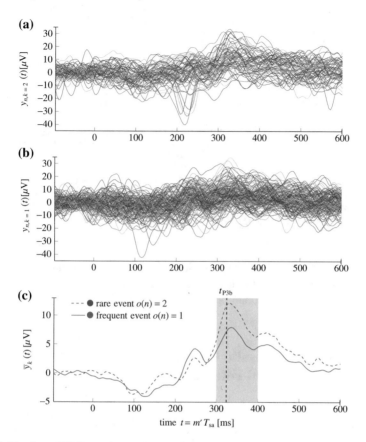

**Fig. 1.2** Epochs and ERP waveforms of the EEG at electrode Pz. **a** $N_{k=2} = 65$ epochs measured in response to the rare event $k = 2$. **b** $N_{k=1} = 132$ epochs measured in response to the frequent event $k = 1$. **c** Grand-average ERP waves (1.3) for the frequent (——) and rare event (- - -). The P3b ROI from 300 to 400 ms after stimulus presentation is highlighted in *gray* and (- - -) indicates the time index $t_{P3b}$ (1.4) of the maximum difference between the two waves

with $m' \in [75, 100]$ (i.e., $\mathcal{T}_{P3b} = [300, 400]$ ms) and $|\cdot|$ indicating the absolute value of the difference. This ERP component latency is shown as (- - -) in Fig. 1.2c. An integration procedure following

$$y(n) = \frac{1}{9} \sum_{\mu'=-4}^{+4} y_n(m'_{P3b} + \mu') \tag{1.5}$$

attenuates high frequency noise (Luck 2014) and yields the ERP amplitudes $y(n)$ which can finally be used for further trial-by-trial analyses. Figure 1.3 shows $y(n)$ (——) calculated after (1.5) along with the stimulus sequence, which is depicted by (●) signifying a frequent event ($k = 1$) and by (●) signifying a rare event ($k = 2$).

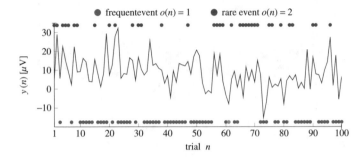

**Fig. 1.3** P3b amplitudes $y(n)$ ((——), (1.5)) for $n \in \{1, \dots, 100\}$ derived from 100 of the epochs depicted in Fig. 1.2. The stimulus sequence is indicated above and below (——) by (●) signifying a frequent event ($k = 1$) and by (●) signifying a rare event ($k = 2$). A trial during which the wrong response was selected is marked by (●)

These symbols will be used throughout this work to denote frequent and rare events. In contrast to Fig. 1.2a, b only trials $n \in \{1, \dots, 100\}$ are shown to permit a clear representation of the corresponding stimulus sequence.

## 1.2   Signal-to-Noise Ratio Estimation for Event-Related Potentials

The signal-to-noise (power) ratio (SNR) of the EEG plays a key role in ERP research: It defines how many trials are necessary for meaningful ERP estimates (Luck 2004) as well as reliable model selection (Penny 2012). The first SNR estimation scheme for ERP data was proposed by Schimmel (1967) just 2 years after the discovery of the P300 (Sutton et al. 1965). While this scheme was employed in the following years (Schimmel et al. 1974), nowadays it is considered "conservative" (Viola et al. 2011). Möcks et al. (1988) proposed an alternative scheme, which is still used as the basis for current developments (beim Graben 2001; Paukkunen et al. 2010) in contrast to other approaches which were proposed in the meantime (Coppola et al. 1978; Basar 1980; Raz et al. 1988; Puce et al. 1994).

Despite the importance of the SNR, it is virtually never reported in ERP studies. In this work, SNRs are calculated following the approach from Möcks et al. (1988), which is now briefly described. For a single subject, the SNR is calculated for each type of event separately and subsequently averaged. To this end, the ERP amplitudes $y(n)$ have to be separated according to event type $k$ as for the ERP waveforms in Sect. 1.1, yielding $y_k(n)$, with $n \in \{1, \dots, N_k\}$ and $N_k$ as the total number of trials in which event $k$ was elicited. The signal model assumes $y_k(n)$ to be composed of the clean signal $s_k$ and additive noise $\epsilon_k(n)$ with variance $\sigma_{\epsilon_k}^2$ (beim Graben 2001), yielding

$$y_k(n) = s_k + \epsilon_k(n). \tag{1.6}$$

Note that $s_k$ is assumed to be constant over trials and $\epsilon_k(n)$ is assumed to be stationary ergodic noise (beim Graben 2001). Although these assumptions are not met for real ERP amplitudes, they are accepted as useful simplifications for SNR estimation (Möcks et al. 1988). The SNR for event $k$ is defined as (beim Graben 2001)

$$\text{SNR}_k = \frac{P_{s_k}}{P_{\epsilon_k}}, \tag{1.7}$$

which is the ratio of the signal power

$$P_{s_k} = s_k^2 \tag{1.8}$$

over the noise power

$$P_{\epsilon_k} = \sigma_{\epsilon_k}^2. \tag{1.9}$$

The power of the measured signal is as follows (beim Graben 2001)

$$P_{y_k} = \overline{y_k^2} = \frac{1}{N_k} \sum_{n=1}^{N_k} y_k^2(n) \tag{1.10}$$

and, assuming statistical independence between the signal and the noise, is composed of the power of the clean signal and noise according to

$$P_{y_k} = P_{s_k} + \frac{1}{N_k} P_{\epsilon_k}. \tag{1.11}$$

Note that the noise power $P_{\epsilon_k}$ is scaled by $\frac{1}{N_k}$ as the power of the noise left in $P_{y_k}$ is attenuated by the factor $N_k$ (Möcks et al. 1988; Paukkunen et al. 2010; Czanner et al. 2015). As $P_{s_k}$ and $P_{\epsilon_k}$ are not directly measurable, they have to be estimated from $y_k(n)$. Möcks et al. (1988) propose the noise power estimate

$$\widehat{P}_{\epsilon_k} = \frac{1}{N_k - 1} \sum_{n=1}^{N_k} (y_k(n) - \bar{y}_k)^2, \tag{1.12}$$

with

$$\bar{y}_k = \frac{1}{N_k} \sum_{n=1}^{N_k} y_k(n), \tag{1.13}$$

and solve (1.11) for the signal power estimate

$$\widehat{P}_{s_k} = P_{y_k} - \frac{1}{N_k} \widehat{P}_{\epsilon_k}. \tag{1.14}$$

Consequently, the signal-to-noise ratio estimate $\widehat{\text{SNR}}_k$ for event $k$ follows

$$\widehat{\text{SNR}}_k = \frac{\widehat{P}_{s_k}}{\widehat{P}_{\epsilon_k}},$$

(1.15)

or in [dB]

$$\widehat{\text{SNR}}_k \, [\text{dB}] = 10 \log_{10} \frac{\widehat{P}_{s_k}}{\widehat{P}_{\epsilon_k}}.$$

(1.16)

Note that the ERP amplitudes in response to event type $k$ are not constant over trials (Squires et al. 1976; Mars et al. 2008; Kolossa et al. 2013; Ostwald et al. 2012; Lieder et al. 2013; Kolossa et al. 2015). Hence, the variance of the clean signal, which is contained in the measured signal, is modeled to be part of the noise power. Thus, this estimation scheme systematically overestimates the noise power and, consequently, underestimates the signal-to-noise ratio. The weighted average estimate $\widehat{\text{SNR}}$ over all types of events $\mathcal{K}$ is

$$\widehat{\text{SNR}} \, [\text{dB}] = \frac{1}{\sum_{k \in \mathcal{K}} N_k} \sum_{k \in \mathcal{K}} N_k \cdot \widehat{\text{SNR}}_k \, [\text{dB}].$$

(1.17)

For the EEG data presented in the example in Sect. 1.1, the stimulus-specific estimated signal-to-noise ratios (1.16) are $\widehat{\text{SNR}}_{k=1} \, [\text{dB}] = 2.27$ dB for the frequent event and $\widehat{\text{SNR}}_{k=2} \, [\text{dB}] = 4.28$ dB for the rare event. The average SNR estimate (1.17) is $\widehat{\text{SNR}} \, [\text{dB}] = 2.93$ dB.

## 1.3   Circularity in Data Analyses

As detailed in Sect. 1.1, the definition of the region of interest (ROI) and the search criterion for the ERP component latencies are crucial steps for deriving the ERP amplitudes. When defining the ROI as well as the search criterion, it is important to avoid the so-called *circularity* or *double dipping*, i.e., data selection based on the same features or statistics which are part of subsequent analyses. Vul et al. (2009) sparked off a heated discussion about the validity of published findings which exceeded theoretical limits due to circularity in data analyses (Barrett 2009; Lazar 2009; Lieberman et al. 2009; Lindquist and Gelman 2009; Yarkoni 2009; Fiedler 2011). A consensus was reached that "Inference based on nonindependent selective analyses is not statistically sound and is never acceptable" (Kriegeskorte et al. 2010), which had a measurable impact on the results published within the neuroimaging community (Vul and Pashler 2012).

This work follows the policy for noncircular analyses as proposed by Kriegeskorte et al. (2009). Specifically, the electrodes and the time windows of interest are defined purely based on the literature. Within the ROI, the search for the ERP component latencies relies purely on the maximal effect of the experimental design on the ERP waveforms, which is completely independent of the subsequent model-based analyses. The study-specific procedures for the selection of ERP data are detailed in Sects. 3.2 and 4.2, respectively.

The same restrictions applying to data selection hold for model parameter identification as well. Consequently, the same policies are used to prevent circularity. The data used for the model parameter training have to be independent from the data used for model selection. This is achieved using separate data sets for model training and selection, which is done in Chap. 3, or using behavioral data for the model training as in Chap. 4.

## 1.4 Probabilities and Surprise

Figure 1.4 shows how evidence for the coding of probabilistic quantities in the brain is obtained using a framework that relates random variables to neural activities: The observer models under test keep track of the probability distributions over observable and hidden random variables. The probability distributions over *hidden* random variables are called *beliefs*, while the distributions over *observable* random variables are called *predictions*. Computations based on these probability distributions are reflected in *Bayesian updating* and *predictive surprise*, which are called *response functions* (Lieder et al. 2013) and map the probability distributions to neural activities in terms of ERP amplitudes. Note that while the task employed in Chap. 3 is limited to observable random variables, both types of variables are present in the task

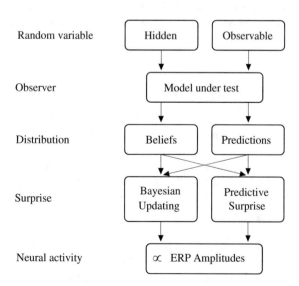

**Fig. 1.4** Hierarchical structure that relates the random variables and observer models to neural activities. There are hidden and observable random variables which are tracked by the observer models in terms of probability distributions. Bayesian updating and predictive surprise are response functions that link the probability distributions to the ERP amplitudes

in Chap. 4. This section defines the random variables, their probability distributions, and how they are related to neural activities via surprise. Note that the actual calculi performed by the observer models under test are detailed later on in Chaps. 3 and 4.

The observable random variables are the $K$ different types of events $k \in \mathcal{K} = \{1, \ldots, K\}$, which are presented to the subjects as stimuli on a trial-by-trial basis. On each trial $n \in \{1, \ldots, N\}$, event $k$ takes the form of an observation $o(n) = k$. The observation probability $P(o(n) = k | \mathbf{o}_1^{n-1}) = P_k(n)$ is the probability for event $k$ to be observed on trial $n$ given a sequence $\mathbf{o}_1^{n-1} = (o(1), o(2), \ldots, o(n-1))$ of $n-1$ previous observations. Consequently, $P(o(n+1) = k | \mathbf{o}_1^n) = P_k(n+1)$ is the probability for event $k$ to be observed on trial $n+1$ given a sequence $\mathbf{o}_1^n = (o(1), o(2), \ldots, o(n))$ of $n$ previous observations. Taking $P_k(n)$ and $P_k(n+1)$ for all $k \in \mathcal{K}$ yields, the probability distributions $P_\mathcal{K}(n)$ and $P_\mathcal{K}(n+1)$, respectively. These distributions are called *predictions*, as they are concerned with *observable* random variables.

Let the hidden random variables be the $U$ different types of states $u \in \mathcal{U} = \{1, \ldots, U\}$, which may represent a hidden state $q = u$. The probability for state $q$ being $u$ after a sequence $\mathbf{o}_1^{n-1}$ of $n-1$ observations is $P(q(n-1) = u | \mathbf{o}_1^{n-1}) = P_u(n-1)$, while it is $P(q(n) = u | \mathbf{o}_1^n) = P_u(n)$ after $n$ observations. The probability distributions $P_\mathcal{U}(n-1)$ and $P_\mathcal{U}(n)$ combine $P_u(n-1)$ and $P_u(n)$ for all $u \in \mathcal{U}$, respectively. These distributions are called *beliefs*, as they are concerned with *hidden* random variables.

The Bayesian brain hypothesis states that such probability distributions are encoded in the human brain (Knill and Pouget 2004; Friston 2005; Doya et al. 2007; Gold and Shadlen 2007; Kopp 2008). Accordingly, observations $o(n)$ trigger computations based on them. These computations may be their evolution over time as the information contained in observation $o(n)$ is incorporated, or, alternatively, a single probability can be taken from the distribution to calculate the information contained in the observation. It is assumed that some ERPs at distinct latencies are measurable traces of these computations (Donchin 1981; Mars et al. 2008; Ostwald et al. 2012; O'Reilly et al. 2013). The following section introduces predictive surprise $I_P$ and Bayesian updating $I_B$, which is instantiated either as postdictive surprise or Bayesian surprise (Kolossa et al. 2015).

### 1.4.1 Bayesian Updating

Bayesian updating subsumes postdictive surprise and Bayesian surprise. Both kinds of surprise reflect the change in probability distributions induced by observations, which is measured as the Kullback–Leibler divergence $D_{KL}$ (see, e.g., Itti and Baldi 2009; Baldi and Itti 2010). The difference, however, lies in the nature of the random variables which are observable or hidden, respectively.

**Postdictive Surprise**

For the probability distributions over the *observable* random variables $k \in \mathcal{K}$, the degree of updating from $P_\mathcal{K}(n)$ to $P_\mathcal{K}(n+1)$ induced by the observation $o(n)$ is referred to as *postdictive* surprise $I_B(n)$ (Kolossa et al. 2015). It is calculated via

$$I_B(n) = D_{KL}(P_{\mathcal{K}}(n) \parallel P_{\mathcal{K}}(n+1)) = \sum_{k \in \mathcal{K}} P_k(n) \log \left( \frac{P_k(n)}{P_k(n+1)} \right). \tag{1.18}$$

**Bayesian Surprise**

For the probability distributions over the *hidden* random variables $u \in \mathcal{U}$ the Kullback-Leibler divergence between $P_{\mathcal{U}}(n-1)$ and $P_{\mathcal{U}}(n)$ is called *Bayesian* surprise $I_B(n)$ (Ostwald et al. 2012)

$$I_B(n) = D_{KL}(P_{\mathcal{U}}(n-1) \parallel P_{\mathcal{U}}(n)) = \sum_{u \in \mathcal{U}} P_u(n-1) \log \left( \frac{P_u(n-1)}{P_u(n)} \right). \tag{1.19}$$

## *1.4.2 Predictive Surprise*

While *postdictive* surprise (1.18) is based on the distributions $P_{\mathcal{K}}(n)$ and $P_{\mathcal{K}}(n+1)$ over *all* events $k \in \mathcal{K}$, *predictive* surprise $I_P(n)$ is based on the *single* probability $P_{k=o(n)}(n)$ taken from the distribution $P_{\mathcal{K}}(n)$, which corresponds to the event $k$ observed on trial $n$ (Mars et al. 2008; Kolossa et al. 2013). Formally, predictive surprise is the information content of $o(n)$ and is calculated according to (Shannon and Weaver 1948; Strange et al. 2005)

$$I_P(n) = -\log_2 P_{k=o(n)}(n). \tag{1.20}$$

If an event with a high probability is observed, it has a small information content and is thus not surprising, while the observation of an event with a low probability contains a lot of information and is very surprising. Thus, predictive surprise is a measure of how unexpected the observation $o(n)$ is (O'Reilly et al. 2013). It is not possible to calculate predictive surprise (1.20) based on the hidden states, as they are not observed at any time, and thus no single probability $P_{u=q(n)}(n)$ can be chosen from the distribution $P_{\mathcal{U}}(n)$. In order to model an information theoretic transformation of the probability distribution over the hidden random variables, the entropy $I_H(n)$ is calculated following

$$I_H(n) = -\sum_{u \in \mathcal{U}} P_u(n) \log_2 P_u(n). \tag{1.21}$$

Entropy is the average predictive surprise and maximal for equal probabilities $P_u(n) = \frac{1}{U}$ for all $u \in \mathcal{U}$.

## 1.5   Probability Weighting Functions

This work investigates if and how the coding and computing of probabilities P are reflected in event-related brain potentials. But how are probabilities perceived by humans? This question has been addressed in different fields such as economics (Kahneman and Tversky 1979; Tversky and Kahneman 1992; Fox and Poldrack 2009), decision behavior (Gonzalez and Wu 1999; Hsu et al. 2009), cognition (Hilbert 2012), as well as frequency and probability estimation in visual tasks (Varey et al. 1990). The results are concordant across fields (Zhang and Maloney 2012), even on the neuronal level (Tobler et al. 2008; Takahashi et al. 2010): Human estimates of probabilities differ systematically from their objective values. This variation is commonly described as an (inverse) S-shaped probability weighting function $w(P)$, which will be used in this work (Kahneman and Tversky 1979; Prelec 1998; Gonzalez and Wu 1999; Fennell and Baddeley 2012; Zhang and Maloney 2012; Cavagnaro et al. 2013). Figure 1.5 shows the effect of (inverse) S-shaped probability weighting by showing probabilities (——) and the corresponding weighted probabilities (- - -): Low probabilities are overestimated and high probabilities are underestimated. Zhang and Maloney (2012) propose the weighting function

$$\mathrm{Lo}(w(P)) = \zeta \mathrm{Lo}(P) + (1-\zeta)\mathrm{Lo}(P_0), \tag{1.22}$$

with crossover point $P_0$ and the log-odds function (Barnard 1949)

$$\mathrm{Lo}(P) = \log\left(\frac{P}{1-P}\right). \tag{1.23}$$

**Fig. 1.5** Comparison of objective probabilities ($\zeta = 1$, (——)) with inverse S-shaped weighted probabilities ($\zeta = 0.65$, (- - -)) after (1.24) with crossover point $P_0 = 0.5$. Probabilities lower than 0.5 are overestimated, and probabilities higher than 0.5 are underestimated, which is represented by the horizontal line (······) at $w(P) = 0.5$

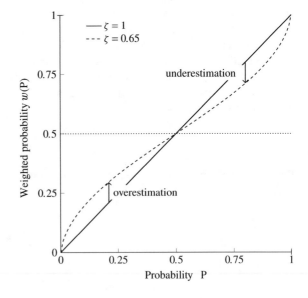

Inserting (1.23) in (1.22) and solving for $w(P)$ yields

$$w(P) = \frac{\left(\frac{P}{1-P}\right)^{\zeta} \cdot \left(\frac{P_0}{1-P_0}\right)^{1-\zeta}}{1 + \left(\frac{P}{1-P}\right)^{\zeta} \cdot \left(\frac{P_0}{1-P_0}\right)^{1-\zeta}}. \tag{1.24}$$

The shape parameter $\zeta$ controls the curvature of the weighting function, with $\zeta = 1$ yielding the identity $w(P) = P$, $\zeta > 1$ causing S-shaped weighting, and $\zeta < 1$ leading to inverse S-shaped weighting. The crossover point $P_0$ is the single probability which is not altered by the weighting function ($w(P_0) = P_0$). By setting the crossover point $P_0 = 0.5$ the weighting function is point-symmetric with respect to $P_0$, i.e.,

$$1 - w(P) = w(1 - P). \tag{1.25}$$

This property of the probability weighting function (1.24) is illustrated in Fig. 1.5. In this work, weighted probabilities are denoted as

$$P^{(w)} = w(P), \tag{1.26}$$

for sake of clarity. The effect of probability weighting on probabilistic inference and behavior will be investigated in Chap. 4.

# Chapter 2
# Introduction to Model Estimation and Selection Methods

When conducting interdisciplinary research, the employed methods may not be common knowledge in all involved fields. This chapter serves to make this work accessible to a wide audience by describing in detail the methods used for model estimation and selection, which are the fundamental concepts of model-based analyses (Mars et al. 2012). The first sections aim at making the reader accustomed to the terminology and concepts underlying the different kinds of hypotheses and models classically used in (neuro-)psychology. After showing cases where classical methods are inadequate for model selection, state of the art in Bayesian methods are presented.

This chapter is structured as follows: It starts with a simple exemplary study which is used to illustrate different types of hypotheses and models as well as statistical methods for comparing them. The presented methods are not an exhaustive list, as this would be beyond the scope of this work. Next, a detailed description of hierarchical linear models and the parametric empirical Bayes (PEB) schemes employed for all analyses in this work is given. A tutorial with an example experiment that inspired Kolossa et al. (2016) concludes this chapter.

A short note on notation: In the following, not emphasized letters refer to scalars, bold lowercase letters to (column) vectors, bold capital letters to matrices, and the superscript $[\;]^T$ denotes the transpose.

## 2.1 An Example Study

This section describes a fictitious study which exemplifies the different kinds of hypotheses and models in the following sections. Has learning for an exam any effect on the achieved result? In order to answer this question, a study is carried out in which a test subject takes a total of $N$ exams, each of which is indexed with a trial number $n \in \{1, \ldots, N\}$. For each trial, the time spent learning and the achieved points are recorded, forming the predictors $x(n)$ and dependent variables $y(n)$, respectively. Note that the dependent variable is simply the measured data, which are the ERP sequences in the rest of this work. The examples in this chapter are the only instances

© Springer International Publishing Switzerland 2016
A. Kolossa, *Computational Modeling of Neural Activities for Statistical Inference*, DOI 10.1007/978-3-319-32285-8_2

where the $y(n)$ contains data other than the ERP sequence. The single-level models described in the following section represent a single subject. Multiple-level models representing more than one subject at once are presented subsequently.

## 2.2   Classical Single-Level Models

Classical hypothesis testing as used in psychological research relies on $p$-values, which will be explained shortly (Fisher 1926; Neyman and Pearson 1933). In some cases, these tests do not give the answers the researchers actually seek, but are still used lacking more sophisticated methods (Cohen 1994; Goodman 1999a). More useful methods have been proposed, with Bayes factors being the most advantageous (see, e.g., Kass and Raftery 1995; Goodman 1999b; Friston et al. 2002; Hoijtink 2012; Penny 2012). This section shortly comments on evaluation methods for informative hypotheses before going into detail about methods for trial-by-trial models which model each individual data point. It closes with showing how Bayesian evaluation schemes work and why they were used in this work.

### 2.2.1   The Null and Informative Hypotheses

To test the hypothesis that learning for exams has *any* effect on the achieved points (see Sect. 2.1), the subject does not prepare for one group of exams while he/she does prepare for the other group. This yields two experimental conditions $c \in C = \{1, 2\}$ with $N_c$ trials each and the total number of trials $N = \sum_{c \in C} N_c$. The relation of the points achieved in the two conditions is of interest. The null hypothesis $H_0$ constitutes that learning does not have any effect, i.e., that there is no difference in the condition-specific mean points $\mu_c$. It is formalized via the equality

$$\mu_1 = \mu_2, \tag{2.1}$$

with $\mu_1$ as the mean points for tests taken without learning and $\mu_2$ as the mean points for tests taken with learning. Classically, the null hypothesis is tested using $p$-values (Rutherford 2001; Weiss 2006). They represent the probability of getting the observed (or more extreme) data in the absence of any effect, i.e., if $H_0$ was true, but *not* the probability of $H_0$ being true (Biau et al. 2010). If the $p$-value is sufficiently small, the null hypothesis is rejected. This is sometimes confused with the likelihood that a specific effect of interest is present, which is not valid because $H_0$ can be false due to *any* effect (Cohen 1994). Ronald Fisher (1926) proposed an arbitrary boundary saying "We shall not often be astray if we draw a conventional line at 0.05...", which everyone followed suit. Soon $p = 0.05$ was established as *the* significance boundary to reject the null hypothesis, which is contested because

"... surely, God loves the 0.06 nearly as much as the 0.05." (Rosnow and Rosenthal 1989) and led to a bias of $p$-values just below 0.05 in publications (Masicampo and Lalande 2012).

This traditional null hypothesis testing is nowadays challenged as outdated "20th century thinking" (Osborne 2010). The main argument is that the null hypothesis is never true as "the probability that an effect is exactly zero is itself zero." (Friston and Penny 2003). Consequently, by increasing the amount of data, one can always reject the null hypothesis with $p \leq 0.05$ (Cohen 1994; Van de Schoot et al. 2011; Friston 2012), which has been exaggerated to the point that the collection of data is unnecessary for rejecting the null hypothesis (Royall 1997). See Wagenmakers (2007) for an overview on the critiques against $p$-values and Wainer (1999) for cases where they are useful.

Hoijtink et al. (2008) propose the evaluation of informative hypotheses using Bayes factors based on accuracy and complexity terms. Bayes factors can be used to select between competing informative hypotheses like

$$\mu_1 > \mu_2, \tag{2.2}$$

which states that $\mu_1$ is larger than $\mu_2$, or

$$\mu_1 < \mu_2, \tag{2.3}$$

which states that $\mu_1$ is smaller than $\mu_2$. While these methods accommodate more sophisticated hypotheses than classical approaches, they are still not quantitatively incorporating the single-trial predictors $x(n)$ and are not capable to select between *trial-by-trial* models as proposed in this work.

## 2.2.2 The General Linear Model

The general linear model accommodates trial-by-trial models, which incorporate the quantitative influence of learning for each single exam as well as an error term. As defined in Sect. 2.1, the amount of learning for an exam $n \in \{1, ..., N_c\}$ is quantified as predictor $x(n)$ (for ease of presentation $x(n) \in \mathbb{R}$ is assumed in the following). The achieved points of that exam $y(n)$ (for ease of presentation $y(n) \in \mathbb{R}$ is assumed in the following) are modeled via the single-level general linear model (GLM)

$$y(n) = x(n)\theta + \epsilon(n), \tag{2.4}$$

with the exam-independent unknown parameter $\theta$, which parameterizes the influence of learning and $\epsilon(n)$ as the exam-specific error which encompasses all deviations from this model. These may be due to the student having a particularly bad or good day, the test being especially difficult or easy, or any other influences on the achieved grade. The term $x(n)\theta$ can be interpreted as the estimate $\widehat{s}(n)$ of the clean data $s(n)$, yielding

$$y(n) = \widehat{s}(n) + \epsilon(n) = x(n)\theta + \epsilon(n). \tag{2.5}$$

The parameter $\theta$ is classically estimated by minimizing the error $\epsilon(n)$, using linear regression (Fahrmeir and Tutz 1994). While (2.5) models only one data point, it can be expressed for all exams simultaneously in matrix notation as

$$\mathbf{y} = \widehat{\mathbf{s}} + \epsilon = \mathbf{x}\theta + \epsilon, \tag{2.6}$$

with data vector $\mathbf{y} = [y(n=1), ..., y(n=N)]^T \in \mathbb{R}^N$, clean data estimate $\widehat{\mathbf{s}} = [\widehat{s}(n = 1), ..., \widehat{s}(n = N)]^T \in \mathbb{R}^N$, design matrix (here a vector) $\mathbf{x} = [x(n = 1), ..., x(n = N)]^T \in \mathbb{R}^N$, and error vector $\epsilon = [\epsilon(n=1), ..., \epsilon(n = N)]^T \in \mathbb{R}^N$. Examining (2.6) makes the exam-independent nature of the parameter $\theta$ apparent.

For more complex models, the framework of the GLM accommodates not only one single but $R$ different predictors $x_r(n)$ with $r \in \mathcal{R} = \{1, ..., R\}$. These predictors can, e.g., be influences like social status, private lessons, gender, etc. Consequently, $R$ model parameters $\theta_r$ represent the weights of the predictors, giving the clean data estimate $\widehat{s}(n)$ the form

$$\widehat{s}(n) = x_{r=1}(n)\theta_{r=1} + ... + x_{r=R}(n)\theta_{r=R} \tag{2.7}$$

$$= \begin{bmatrix} x_{r=1}(n) & \cdots & x_{r=R}(n) \end{bmatrix} \begin{bmatrix} \theta_{r=1} \\ \vdots \\ \theta_{r=R} \end{bmatrix}. \tag{2.8}$$

Modeling all trials at once yields

$$\begin{bmatrix} \widehat{s}(n=1) \\ \vdots \\ \widehat{s}(n=N) \end{bmatrix} = \begin{bmatrix} x_{r=1}(n=1) & \cdots & x_{r=R}(n=1) \\ \vdots & \ddots & \vdots \\ x_{r=1}(n=N) & \cdots & x_{r=R}(n=N) \end{bmatrix} \begin{bmatrix} \theta_{r=1} \\ \vdots \\ \theta_{r=R} \end{bmatrix} \tag{2.9}$$

$$\widehat{\mathbf{s}} = \mathbf{X}\theta, \tag{2.10}$$

which again results in the model of the measured data as in (2.6)

$$\mathbf{y} = \widehat{\mathbf{s}} + \epsilon = \mathbf{X}\theta + \epsilon, \tag{2.11}$$

with the design matrix $\mathbf{X} \in \mathbb{R}^{N \times R}$ and vector $\theta \in \mathbb{R}^R$. A common type of model consists of $R = 2$ predictors with one being a constant $x_1(n) = 1$, which models an intercept (i.e., $\theta_1$ are the points the student would get without any preparation). Consequently, $x_2(n)$ is the amount of time spent learning for test $n$, while $\theta_2$ is the influence of time spent learning on the achieved points.

In contrast to classical or informative hypotheses, competing models $m \in \mathcal{M}$, with $m$ as model index and $\mathcal{M}$ as model space, are not specified by different inequality constraints of means $\mu_c$, but by different predictors in the design matrices $\mathbf{X}$ (e.g., $m =$ time spent learning versus $m =$ time spent playing video games). Note that if the design matrix of one model consists of multiple predictors $x_r$, the estimated parameters $\theta_r$ allow for inference regarding the influence of the specific predictors (Friston et al. 2002). Methods for parameter estimation and model selection are described in Sect. 2.4.

## 2.3 Hierarchical Multiple-Level Models

A single-level GLM (2.11) can be extended by additional levels which allow for the parameters of a lower level themselves to be modeled by a higher level. In these multiple-level GLMs, the first-level models the data $\mathbf{y}$ as a linear combination of predictors $\mathbf{X}^{(1)}$ weighted by parameters $\theta^{(1)}$, and an additive error $\epsilon^{(1)}$, which is exactly (2.11) with an additional superscript $(\ )^{(1)}$ indicating the first level (Kiebel and Holmes 2003). The second level sets priority on the first-level parameters by modeling them as consisting of a design matrix $\mathbf{X}^{(2)}$, parameters $\theta^{(2)}$, and errors $\epsilon^{(2)}$ (Friston and Penny 2003). The second level parameters can again be modeled by a third level, which consists of a design matrix $\mathbf{X}^{(3)}$, parameters $\theta^{(3)}$, and errors $\epsilon^{(3)}$ (Friston et al. 2002). This section elaborates on multiple-level GLMs within the scope to which they are applied in this work. Readers interested in a more generalized introduction to linear hierarchical models are referred to Fahrmeir and Tutz (1994).

### 2.3.1 The First Level

The first level of a multiple-level GLM is the same as the single-level GLM (2.11)

$$
\begin{bmatrix} y(n=1) \\ \vdots \\ y(n=N) \end{bmatrix} = \begin{bmatrix} x_{r=1}^{(1)}(n=1) & \cdots & x_{r=R}^{(1)}(n=1) \\ \vdots & \ddots & \vdots \\ x_{r=1}^{(1)}(n=N) & \cdots & x_{r=R}^{(1)}(n=N) \end{bmatrix} \begin{bmatrix} \theta_{r=1}^{(1)} \\ \vdots \\ \theta_{r=R}^{(1)} \end{bmatrix} + \begin{bmatrix} \epsilon^{(1)}(n=1) \\ \vdots \\ \epsilon^{(1)}(n=N) \end{bmatrix}
$$
(2.12)

or, in matrix notation,

$$
\mathbf{y} = \mathbf{X}^{(1)}\theta^{(1)} + \epsilon^{(1)},
$$
(2.13)

with data vector $\mathbf{y} \in \mathbb{R}^N$, design matrix $\mathbf{X}^{(1)} \in \mathbb{R}^{N \times R}$, parameter vector $\theta^{(1)} \in \mathbb{R}^R$, and error vector $\epsilon^{(1)} \in \mathbb{R}^N$.

### 2.3.2   The Second Level

The second level models the first-level parameters $\boldsymbol{\theta}^{(1)}$ following

$$\boldsymbol{\theta}^{(1)} = \mathbf{X}^{(2)}\boldsymbol{\theta}^{(2)} + \boldsymbol{\epsilon}^{(2)}, \tag{2.14}$$

with design matrix $\mathbf{X}^{(2)} \in \mathbb{R}^{R \times R}$, parameters $\boldsymbol{\theta}^{(2)} \in \mathbb{R}^R$, and error $\boldsymbol{\epsilon}^{(2)} \in \mathbb{R}^R$ (Penny et al. 2003). If the second level design matrix $\mathbf{X}^{(2)}$ is chosen to be all zeros $\mathbf{X}^{(2)} = \mathbf{0} \in \mathbb{R}^{R \times R}$, an unconstrained prior is set on the first-level parameters $\boldsymbol{\theta}^{(1)}$ (Friston et al. 2007)

$$\boldsymbol{\theta}^{(1)} = \boldsymbol{\epsilon}^{(2)}, \tag{2.15}$$

which allows for single-level Bayesian inference (Ostwald et al. 2012). This two-level GLM

$$\mathbf{y} = \mathbf{X}^{(1)}\boldsymbol{\theta}^{(1)} + \boldsymbol{\epsilon}^{(1)}$$
$$\boldsymbol{\theta}^{(1)} = \boldsymbol{\epsilon}^{(2)}. \tag{2.16}$$

is used for Bayesian model estimation throughout this work.

### 2.3.3   The Third Level

For studies with $L$ different subjects $\ell \in \{1, ..., L\}$, the second level can be used to facilitate a mixed effects analysis (Friston et al. 2002). In this case, each subject is modeled with an individual set of first-level parameters $\boldsymbol{\theta}_\ell^{(1)}$ which are shrunk towards subject-independent second level group parameters $\boldsymbol{\theta}^{(2)}$. These parameters are consequently modeled by a third level

$$\boldsymbol{\theta}^{(2)} = \mathbf{X}^{(3)}\boldsymbol{\theta}^{(3)} + \boldsymbol{\epsilon}^{(3)}, \tag{2.17}$$

with design matrix $\mathbf{X}^{(3)} \in \mathbb{R}^{R \times R}$, parameters $\boldsymbol{\theta}^{(3)} \in \mathbb{R}^R$, and error $\boldsymbol{\epsilon}^{(3)} \in \mathbb{R}^R$. Such three-level GLMs are used in the studies by Mars et al. (2008) and Kolossa et al. (2013), where the data from all $L$ subjects is modeled simultaneously. These studies use an unconstrained prior on the second level parameters $\boldsymbol{\theta}^{(2)}$ by employing an all-zero third-level design matrix $\mathbf{X}^{(3)} = \mathbf{0} \in \mathbb{R}^{R \times R}$. The detailed composition of the vectors and matrices is as follows: The subject-specific data vectors $\mathbf{y}_\ell \in \mathbb{R}^{N_\ell}$ with $N_\ell$ as the number of trials of a subject $\ell$, design matrices $\mathbf{X}_\ell \in \mathbb{R}^{N_\ell \times R}$, first-level parameters $\boldsymbol{\theta}_\ell^{(1)} \in \mathbb{R}^R$, and error vectors $\boldsymbol{\epsilon}_\ell^{(1)} \in \mathbb{R}^{N_\ell}$ are augmented to yield the first level of the GLM (Friston et al. 2002)

$$
\begin{bmatrix} \mathbf{y}_{\ell=1} \\ \vdots \\ \mathbf{y}_{\ell=L} \end{bmatrix} = \begin{bmatrix} \mathbf{X}^{(1)}_{\ell=1} & \mathbf{0}_{N_{\ell=1} \times R} & \cdots & \mathbf{0}_{N_{\ell=1} \times R} \\ \mathbf{0}_{N_{\ell=2} \times R} & \mathbf{X}^{(1)}_{\ell=2} & \ddots & \vdots \\ \vdots & \ddots & \ddots & \mathbf{0}_{N_{\ell=L-1} \times R} \\ \mathbf{0}_{N_{\ell=L} \times R} & \cdots & \mathbf{0}_{N_{\ell=L} \times R} & \mathbf{X}^{(1)}_{\ell=L} \end{bmatrix} \begin{bmatrix} \boldsymbol{\theta}^{(1)}_{\ell=1} \\ \vdots \\ \boldsymbol{\theta}^{(1)}_{\ell=L} \end{bmatrix} + \begin{bmatrix} \boldsymbol{\epsilon}^{(1)}_{\ell=1} \\ \vdots \\ \boldsymbol{\epsilon}^{(1)}_{\ell=L} \end{bmatrix}, \quad (2.18)
$$

with all-zero matrices $\mathbf{0}_{N_\ell \times R} \in \mathbb{R}^{N_\ell \times R}$ specifying inter-subject independence of the parameters $\boldsymbol{\theta}^{(1)}_\ell$. Note how this model allows for unbalanced data sets, i.e., variable $N_\ell$ over subjects. The first level (2.18) is written in condensed form as

$$
\mathbf{y} = \mathbf{X}^{(1)} \boldsymbol{\theta}^{(1)} + \boldsymbol{\epsilon}^{(1)}, \quad (2.19)
$$

with data vector $\mathbf{y} \in \mathbb{R}^N$, where $N = \sum_{\ell=1}^{L} N_\ell$ is the total number of trials over subjects, design matrix $\mathbf{X}^{(1)} \in \mathbb{R}^{N \times LR}$, parameter vector $\boldsymbol{\theta}^{(1)} \in \mathbb{R}^{LR}$, and error vector $\boldsymbol{\epsilon}^{(1)} \in \mathbb{R}^N$. The second level models the subject-individual first-level parameters $\boldsymbol{\theta}^{(1)}_\ell$ as samples from subject-independent group parameters $\boldsymbol{\theta}^{(2)} \in \mathbb{R}^R$. This requires a second level design matrix $\mathbf{X}^{(2)} \in \mathbb{R}^{LR \times R}$ that consists of $L$ stacks of identity matrices $\mathbf{I}_R \in \mathbb{R}^{R \times R}$, giving the second level the form

$$
\begin{bmatrix} \boldsymbol{\theta}^{(1)}_{\ell=1} \\ \vdots \\ \boldsymbol{\theta}^{(1)}_{\ell=L} \end{bmatrix} = \begin{bmatrix} \mathbf{I}_R \\ \vdots \\ \mathbf{I}_R \end{bmatrix} \boldsymbol{\theta}^{(2)} + \begin{bmatrix} \boldsymbol{\epsilon}^{(2)}_{\ell=1} \\ \vdots \\ \boldsymbol{\epsilon}^{(2)}_{\ell=L} \end{bmatrix}, \quad (2.20)
$$

which is summarized as the standard second level (2.14)

$$
\boldsymbol{\theta}^{(1)} = \mathbf{X}^{(2)} \boldsymbol{\theta}^{(2)} + \boldsymbol{\epsilon}^{(2)}. \quad (2.21)
$$

The third level (2.17) sets an unconstrained prior on the parameters of the second level via an all-zero third-level design matrix $\mathbf{X}^{(3)} = \mathbf{0} \in \mathbb{R}^{R \times R}$, which yields

$$
\boldsymbol{\theta}^{(2)} = \boldsymbol{\epsilon}^{(3)}, \quad (2.22)
$$

and, in summary, the complete three-level GLM

$$
\begin{aligned}
\mathbf{y} &= \mathbf{X}^{(1)} \boldsymbol{\theta}^{(1)} + \boldsymbol{\epsilon}^{(1)} \\
\boldsymbol{\theta}^{(1)} &= \mathbf{X}^{(2)} \boldsymbol{\theta}^{(2)} + \boldsymbol{\epsilon}^{(2)} \\
\boldsymbol{\theta}^{(2)} &= \boldsymbol{\epsilon}^{(3)}.
\end{aligned} \quad (2.23)
$$

## 2.4   Model Estimation and Selection

After having specified the linear models in the previous sections, final results are obtained in two steps: (1) Estimation of the unknown parameters $\theta$, and (2) calculation of the model likelihoods for model selection. This section starts with motivating the use of Bayesian model estimation methods, which is followed by a detailed description of the estimation schemes for the two-level GLM used in this work (see Sect. 2.3.2). It is concluded by instructions for Bayesian model selection. Readers interested in generalized Bayesian estimation schemes for GLMs of any order are referred to Friston et al. (2002).

A well-known method for single-level model estimation and selection is based on the mean squared error (MSE) between the clean data estimates $\widehat{\mathbf{s}}$ (2.5) and the measured data $\mathbf{y}$ (Kleinbaum et al. 2013)

$$\text{MSE}\,(\widehat{\mathbf{s}},\,\mathbf{y}) = \frac{1}{N}\sum_{n=1}^{N}(\widehat{s}(n) - y(n))^{2}. \tag{2.24}$$

(1) For each model $m \in \mathcal{M}$, the parameters $\theta_m$ are optimized by finding the parameters which yield the smallest MSE

$$\theta_{m,\text{opt}} = \arg\min_{\theta_m}\{\text{MSE}\,(\widehat{\mathbf{s}}_m,\,\mathbf{y})\}. \tag{2.25}$$

(2) Now the models employ $\theta_{m,\text{opt}}$ and the model

$$m_{\text{opt}} = \arg\min_{m}\{\text{MSE}(\widehat{\mathbf{s}}_{m,\text{opt}},\,\mathbf{y})\}, \tag{2.26}$$

with the smallest MSE is selected as the best model $m_{\text{opt}}$, as it offers the best fit (or accuracy) of the measured data. A major shortcoming of this comparison scheme is its blindness to the complexity of the models, i.e., the number $R$ of model parameters $\theta_r$, which are used to calculate $\widehat{s}(n)$.

Bayesian evaluation schemes take the complexity of the models into account by employing a penalty factor for complexity, which is often referred to as *Occam's razor* (MacKay 1992). The reason is to choose the least complex model that offers a good explanation of the data (Myung and Pitt 1997). The most vivid factor influencing model complexity is the number $R$ of model parameters. In the context of general linear models, the design matrix of the least complex model consists of only one constant predictor per trial (i.e., $R=1$), yielding an all-one vector $\mathbf{x}^{(1)} = [1...1]^{T} \in \mathbb{R}^{N}$, which is equivalent to the classical null hypothesis. Consequently, the most complex model fits the data perfectly and has as much predictors as trials, i.e., $R=N$, and an identity design matrix $\mathbf{X}^{(1)} = \mathbf{I}_{N} \in \mathbb{R}^{N \times N}$. Obviously, this model is overfitted and offers no theory behind the data-generating process (MacKay 1992; Pitt and Myung 2002). In this work, these shortcomings are addressed by parametric empirical Bayesian (PEB) methods used for parameter estimation and model selection (Friston et al. 2002).

In this framework, competing models are selected based on their log-likelihoods, which are derived by taking a complexity-accuracy trade-off into account (Friston et al. 2007). It is important to note that complexity is not solely based on the number of model parameters, but on other factors like parameter independence as well (see Stephan et al. 2009 for details). This Bayesian framework is proven to be a useful tool for model selection in many fields (Hoeting et al. 1999; Pitt and Myung 2002; Penny et al. 2010) and has been able to significantly advance neuroimaging research (Woolrich 2012).

### 2.4.1   Collapsing and Augmenting the Hierarchical Model

The first step for parameter estimation and evidence calculation via PEB are alterations to the model structure of the two-level GLM (see Sect. 2.3.2). Specifically, the model is first collapsed to a non-hierarchical form and subsequently augmented, so that all parameters appear in the error vector and can be estimated at once using expectation maximization (EM) (Friston et al. 2002). For the two-level GLM (2.13), (2.14)

$$\mathbf{y} = \mathbf{X}^{(1)}\boldsymbol{\theta}^{(1)} + \boldsymbol{\epsilon}^{(1)} \tag{2.27}$$

$$\boldsymbol{\theta}^{(1)} = \mathbf{X}^{(2)}\boldsymbol{\theta}^{(2)} + \boldsymbol{\epsilon}^{(2)} \tag{2.28}$$

the errors on both levels are assumed to be normally distributed

$$\boldsymbol{\epsilon}^{(1)} \sim \mathcal{N}(0, \Sigma_\epsilon^{(1)}) \tag{2.29}$$

$$\boldsymbol{\epsilon}^{(2)} \sim \mathcal{N}(0, \Sigma_\epsilon^{(2)}), \tag{2.30}$$

with zero mean and isotropic error covariance matrices

$$\Sigma_\epsilon^{(1)} = \lambda^{(1)}\mathbf{I}_N \tag{2.31}$$

$$\Sigma_\epsilon^{(2)} = \lambda^{(2)}\mathbf{I}_R. \tag{2.32}$$

The hyper-parameters $\lambda^{(1)}$ and $\lambda^{(2)}$ control the variances at level (1) and (2), respectively (Mars et al. 2008). They are called hyper-parameters because they parameterize the covariance of the errors $\boldsymbol{\epsilon}^{(1)}$ and $\boldsymbol{\epsilon}^{(2)}$ (Friston and Penny 2003; Penny 2012). Note that this section covers only models where there are no hyper-priors on the hyper-parameters (see Friston et al. 2007 for details). Identity matrices $\mathbf{I}_N \in \mathbb{R}^{N \times N}$ and $\mathbf{I}_R \in \mathbb{R}^{R \times R}$ place independence assumptions over trials and parameters, respectively (Ostwald et al. 2012). While there can be multiple covariance components on any level, only one covariance component per level is assumed in this work. Substitution of (2.28) in (2.27) yields the non-hierarchical form

$$\mathbf{y} = \mathbf{X}^{(1)}\boldsymbol{\epsilon}^{(2)} + \mathbf{X}^{(1)}\mathbf{X}^{(2)}\boldsymbol{\theta}^{(2)} + \boldsymbol{\epsilon}^{(1)} \tag{2.33}$$

$$= \begin{bmatrix} \mathbf{X}^{(1)} \ \mathbf{X}^{(1)}\mathbf{X}^{(2)} \end{bmatrix} \begin{bmatrix} \boldsymbol{\epsilon}^{(2)} \\ \boldsymbol{\theta}^{(2)} \end{bmatrix} + \boldsymbol{\epsilon}^{(1)}, \tag{2.34}$$

which is augmented so that the parameters appear in the error vector

$$\begin{bmatrix} \mathbf{y} \\ \mathbf{0}_{R \times 1} \\ \mathbf{0}_{R \times 1} \end{bmatrix} = \begin{bmatrix} \mathbf{X}^{(1)} & \mathbf{X}^{(1)}\mathbf{X}^{(2)} \\ -\mathbf{I}_R & \mathbf{0}_{R \times R} \\ \mathbf{0}_{R \times R} & -\mathbf{I}_R \end{bmatrix} \begin{bmatrix} \boldsymbol{\epsilon}^{(2)} \\ \boldsymbol{\theta}^{(2)} \end{bmatrix} + \begin{bmatrix} \boldsymbol{\epsilon}^{(1)} \\ \boldsymbol{\epsilon}^{(2)} \\ \boldsymbol{\theta}^{(2)} \end{bmatrix}. \tag{2.35}$$

The augmented model can be expressed condensedly as

$$\widetilde{\mathbf{y}} = \widetilde{\mathbf{X}}\,\widetilde{\boldsymbol{\theta}} + \widetilde{\boldsymbol{\epsilon}}. \tag{2.36}$$

This reformulation of the hierarchical model is computationally efficient and allows an instructive form of the EM algorithm (Friston et al. 2002; Friston et al. 2007). The error covariance matrix of the augmented form is assembled following

$$\Sigma_{\widetilde{\epsilon}} = \sum_{i=1}^{2} \lambda_i \mathbf{Q}_i + \Sigma_{\widetilde{\theta}}, \tag{2.37}$$

with $\lambda_i = \lambda^{(i)}$, as each level is modeled with a single covariance component. The matrices $\mathbf{Q}_1 \in \mathbb{R}^{(N+2R) \times (N+2R)}$ and $\mathbf{Q}_2 \in \mathbb{R}^{(N+2R) \times (N+2R)}$ are the augmented forms of the identity matrices in (2.31) and (2.32), more precisely

$$\mathbf{Q}_1 = \begin{bmatrix} \mathbf{I}_N & \mathbf{0}_{N \times R} & \mathbf{0}_{N \times R} \\ \mathbf{0}_{R \times N} & \mathbf{0}_{R \times R} & \mathbf{0}_{R \times R} \\ \mathbf{0}_{R \times N} & \mathbf{0}_{R \times R} & \mathbf{0}_{R \times R} \end{bmatrix} \tag{2.38}$$

and

$$\mathbf{Q}_2 = \begin{bmatrix} \mathbf{0}_{N \times N} & \mathbf{0}_{N \times R} & \mathbf{0}_{N \times R} \\ \mathbf{0}_{R \times R} & \mathbf{I}_R & \mathbf{0}_{R \times R} \\ \mathbf{0}_{R \times N} & \mathbf{0}_{R \times R} & \mathbf{0}_{R \times R} \end{bmatrix}. \tag{2.39}$$

The parameter covariance matrix $\Sigma_{\widetilde{\theta}} \in \mathbb{R}^{(N+2R) \times (N+2R)}$ is of the form

$$\Sigma_{\widetilde{\theta}} = \begin{bmatrix} \mathbf{0}_{N \times N} & \mathbf{0}_{N \times R} & \mathbf{0}_{N \times R} \\ \mathbf{0}_{R \times N} & \mathbf{0}_{R \times R} & \mathbf{0}_{R \times R} \\ \mathbf{0}_{R \times R} & \mathbf{0}_{R \times R} & \Sigma_{\theta}^{(2)} \end{bmatrix}, \tag{2.40}$$

with $\Sigma_{\theta}^{(2)} = e^{16}\mathbf{I}_R \in \mathbb{R}^{R \times R}$ specifying an unconstrained prior on the second-level parameters, with Euler's number $e = \sum_{\kappa=0}^{\infty} \frac{1}{\kappa!}$.

## 2.4.2 Model Parameter Optimization and Likelihood Calculation

Based on the augmented model (2.36), PEB methods commence to compute the conditional *posterior probability* densities of the *parameters* using *empirical* data, coining the term *parametric empirical Bayes* (Friston et al. 2002). In this framework, the parameters $\widetilde{\theta}$ are modeled to be normally distributed random variables $p(\widetilde{\theta}|\mathbf{y}) = \mathcal{N}(\widetilde{\theta}; \mu_{\widetilde{\theta}|y}, \Sigma_{\widetilde{\theta}|y})$ with conditional means $\mu_{\widetilde{\theta}|y}$ and covariance matrix $\Sigma_{\widetilde{\theta}|y}$. This approach is fundamentally different from classical parameter optimization, where the parameters are assumed to be fixed values.

The parameter densities $p(\widetilde{\theta}|\mathbf{y})$ are estimated via expectation maximization (EM). The EM algorithm estimates the parameter densities by maximizing the free energy $F_{\widetilde{\theta}}$, which gives a lower bound approximation of the log-likelihood of the data conditioned on the hyper-parameters (Friston et al. 2002)

$$F_{\widetilde{\theta}} \leq \log p(\mathbf{y}|\boldsymbol{\lambda}) = \log \int p(\widetilde{\theta}, \mathbf{y}|\boldsymbol{\lambda}) d\widetilde{\theta}. \tag{2.41}$$

See Friston et al. (2002, 2007) for a detailed and thorough derivation. For the augmented general linear model (2.36), the free energy is calculated according to (Friston et al. 2007)

$$F_{\widetilde{\theta}} = -\frac{N}{2} \log(2\pi) - \frac{1}{2}(\mathbf{G}\widetilde{\mathbf{y}})^T \Sigma_{\widetilde{\epsilon}}(\mathbf{G}\widetilde{\mathbf{y}}) + \frac{1}{2} \log |\Sigma_{\widetilde{\epsilon}}^{-1}| + \frac{1}{2} \log |\Sigma_{\widetilde{\theta}|y}|, \tag{2.42}$$

with $|\cdot|$ denoting the determinant of a matrix and $\mathbf{G} = \Sigma_{\widetilde{\epsilon}}^{-1} - \Sigma_{\widetilde{\epsilon}}^{-1} \widetilde{\mathbf{X}} \Sigma_{\widetilde{\theta}|y} \widetilde{\mathbf{X}}^T \Sigma_{\widetilde{\epsilon}}^{-1}$. The composition of all the terms in (2.42) will be described shortly during the summary of the EM algorithm. Generally, EM is an iterative algorithm for maximum likelihood estimation of data conditional on unobserved parameters (Neal and Hinton 1998). Readers completely unfamiliar with EM algorithms are referred to Fahrmeir and Tutz (1994) for a general introduction to EM in the context of linear models. In the PEB framework, the EM algorithm alternates between maximizing the free energy with regard to the parameter distribution $p(\widetilde{\theta}|\mathbf{y}) = \mathcal{N}(\widetilde{\theta}; \mu_{\widetilde{\theta}|y}, \Sigma_{\widetilde{\theta}|y})$ in the E-step and with regard to the hyper-parameters $\boldsymbol{\lambda} = [\lambda_1 \ \lambda_2]^T$ in the M-step. Under Gaussian assumptions, the E-step is simply the calculation of the conditional mean and covariance of the model parameters while keeping the hyper-parameters fixed, following (Friston 2002)

$$\Sigma_{\widetilde{\epsilon}} = \Sigma_{\widetilde{\theta}} + \sum_{i=1}^{2} \lambda_i \mathbf{Q}_i \tag{2.43}$$

$$\Sigma_{\widetilde{\theta}|y} = \left(\widetilde{\mathbf{X}}^T \Sigma_{\widetilde{\epsilon}}^{-1} \widetilde{\mathbf{X}}\right)^{-1} \tag{2.44}$$

$$\mu_{\widetilde{\theta}|y} = \Sigma_{\widetilde{\theta}|y} \widetilde{\mathbf{X}}^T \Sigma_{\widetilde{\epsilon}}^{-1} \widetilde{\mathbf{y}}. \tag{2.45}$$

The M-step serves to estimate the error covariances $\Sigma_{\tilde{\epsilon}}$ which rest upon the hyper-parameters $\lambda$, as can be seen in (2.43). They are obtained by maximizing the free energy $F_{\tilde{\theta}}$ while keeping the conditional mean and covariance of the parameters fixed (Friston et al. 2002)

$$\mathbf{G} = \Sigma_{\tilde{\epsilon}}^{-1} - \Sigma_{\tilde{\epsilon}}^{-1} \tilde{\mathbf{X}} \Sigma_{\tilde{\theta}|y} \tilde{\mathbf{X}}^T \Sigma_{\tilde{\epsilon}}^{-1} \tag{2.46}$$

$$\boldsymbol{h} = \begin{bmatrix} \frac{\partial F_{\tilde{\theta}}}{\partial \lambda_1} \\ \frac{\partial F_{\tilde{\theta}}}{\partial \lambda_2} \end{bmatrix} = \begin{bmatrix} -\frac{1}{2}\mathrm{tr}\{\mathbf{GQ}_1\} + \frac{1}{2}\tilde{\mathbf{y}}^T\mathbf{G}^T\mathbf{Q}_1\mathbf{G}\tilde{\mathbf{y}} \\ -\frac{1}{2}\mathrm{tr}\{\mathbf{GQ}_2\} + \frac{1}{2}\tilde{\mathbf{y}}^T\mathbf{G}^T\mathbf{Q}_2\mathbf{G}\tilde{\mathbf{y}} \end{bmatrix} \tag{2.47}$$

$$\mathbf{H} = \begin{bmatrix} \langle -\frac{\partial^2 F_{\tilde{\theta}}}{\partial \lambda_1 \partial \lambda_1} \rangle & \langle -\frac{\partial^2 F_{\tilde{\theta}}}{\partial \lambda_1 \partial \lambda_2} \rangle \\ \langle -\frac{\partial^2 F_{\tilde{\theta}}}{\partial \lambda_2 \partial \lambda_1} \rangle & \langle -\frac{\partial^2 F_{\tilde{\theta}}}{\partial \lambda_2 \partial \lambda_2} \rangle \end{bmatrix} = \begin{bmatrix} \frac{1}{2}\mathrm{tr}\{\mathbf{GQ}_1\mathbf{GQ}_1\} & \frac{1}{2}\mathrm{tr}\{\mathbf{GQ}_1\mathbf{GQ}_2\} \\ \frac{1}{2}\mathrm{tr}\{\mathbf{GQ}_2\mathbf{GQ}_1\} & \frac{1}{2}\mathrm{tr}\{\mathbf{GQ}_2\mathbf{GQ}_2\} \end{bmatrix} \tag{2.48}$$

$$\lambda = \lambda + \mathbf{H}^{-1}\boldsymbol{h} = \lambda + \Delta\lambda. \tag{2.49}$$

The operator $\mathrm{tr}\{\cdot\}$ denotes the trace of a matrix, $\langle \cdot \rangle$ is the expectation operator, $\mathbf{h}$ is a gradient vector, and $\mathbf{H}$ is referred to as Fisher's information matrix (Friston et al. 2002). Note that (2.46) and (2.47) use the first and expected second partial derivatives of the free energy with regard to the hyper-parameters, which are known as *Fisher scoring* (Friston et al. 2002). Steps (2.43)–(2.48) are repeated until convergence, which can be a specific value of $\Delta\lambda$ or some fixed number of iterations. Readers interested in the formal derivation of (2.43)–(2.48) and more generalized applications, such as multiple covariance constraints on any level or higher order models, are referred to Friston et al. (2002).

After convergence of the EM algorithm, the parameter densities are estimated and the free energy $F_{\tilde{\theta}}$ is adjusted to yield the variational free energy $F$ used for model selection (Friston and Penny 2003; Friston et al. 2007)

$$F = \underbrace{-\frac{N}{2}\log(2\pi) - \frac{1}{2}(\mathbf{G}\tilde{\mathbf{y}})^T\Sigma_{\tilde{\epsilon}}(\mathbf{G}\tilde{\mathbf{y}}) + \frac{1}{2}\log|\Sigma_{\tilde{\epsilon}}^{-1}|}_{\text{accuracy term}} + \underbrace{\frac{1}{2}\log|\Sigma_{\tilde{\theta}|y}| + \frac{1}{2}\log|-\mathbf{H}^{-1}|}_{\text{complexity term}}.$$

$$\tag{2.50}$$

The variational free energy is a lower bound approximation of the usually not directly computable log-likelihood $\log(p(\mathbf{y}|m))$, i.e., the logarithm of the probability of the data $\mathbf{y}$ given the model $m \in \mathcal{M} = \{1, ..., M\}$ (Penny et al. 2010). Many approximations to the log-likelihood have been proposed, with the variational free energy $F$ being superior and commonly used in neuroimaging (Beal 2003; Beal and Ghahramani 2003; Friston et al. 2007; Penny 2012).

The only values of further interest besides the variational free energy are the conditional means of the parameters which serve as their point estimates. They can be used for model fitting or comparison of effect sizes, with the latter being only sound if the data and regressors have been normalized (Hoijtink 2012). Using the vector of the conditional means of the augmented model $\mu_{\tilde{\theta}|y} = \left[ \mu_{\epsilon|y}^{(2)} \ \mu_{\theta|y}^{(2)} \right]^T$ from (2.45), the conditional means of the first-level parameters are calculated according to

$$\mu_{\theta|y}^{(1)} = \mathbf{X}^{(2)} \mu_{\theta|y}^{(2)} + \mu_{\epsilon|y}^{(2)}. \tag{2.51}$$

The described methods for parameter estimation and variational free energy calculation are implemented in the spm_PEB.m function of the freely available Statistical Parametric Mapping (SPM8) software (Dempster et al. 1981; Friston et al. 2002, 2007), which was used for model estimation in this work.

### 2.4.3 Model Selection Using Bayes Factors and Posterior Model Probabilities

After parameter estimation and calculation of the variational free energy $F$ (2.50) in the previous chapter, the best model $m$ can be selected. If the choice is solely among two models $\mathcal{M} = \{1, 2\}$, the Bayes factor (BF) is a suitable measure (Kass and Raftery 1995). It is the ratio of the model likelihoods

$$\text{BF}_{1 \to 2} = \frac{p(\mathbf{y}|m=1)}{p(\mathbf{y}|m=2)}, \tag{2.52}$$

whose natural logarithm, log(BF), equals the difference of variational free energy (Penny et al. 2004)

$$\log(\text{BF}_{1 \to 2}) = \log\left(\frac{p(\mathbf{y}|m=1)}{p(\mathbf{y}|m=2)}\right) = F_{m=1} - F_{m=2}. \tag{2.53}$$

Positive values reflect evidence in favor of model 1 over 2. The interpretation of log-Bayes factors is often unintuitive, and their bilateral nature makes the description of selection procedures with $M > 2$ models unnecessarily complex. The interpretation becomes most vivid and independent of the number of evaluated models by computing posterior model probabilities (PMP) $P(m|\mathbf{y})$, which are calculated based on the model likelihoods following Bayes' rule (Penny et al. 2010)

$$P(m|\mathbf{y}) = \frac{p(\mathbf{y}|m)P(m)}{\sum\limits_{\mu \in \mathcal{M}} p(\mathbf{y}|\mu)P(\mu)}, \quad \forall m \in \mathcal{M}, \tag{2.54}$$

with $P(m)$ as the prior probability of model $m$. Assuming equal prior probabilities of $P(m) = \frac{1}{M}, \forall m \in \mathcal{M}$, (2.54) simplifies to

**Table 2.1** Bayes factors BF, log-Bayes factors log(BF), posterior model probabilities P($m$|$\mathbf{y}$), and how to interpret them (Kass and Raftery 1995; Penny et al. 2004)

| BF$_{1\rightarrow2}$ | log (BF$_{1\rightarrow2}$) | P($m$|$\mathbf{y}$) | Significance |
|---|---|---|---|
| 1–3 | 0–1.1 | 0.50–0.75 | Weak |
| 3–20 | 1.1–3 | 0.75–0.95 | Positive |
| 20–150 | 3–5 | 0.95–0.99 | Strong |
| >150 | >5 | >0.99 | Very strong |

The Bayes factor and log-Bayes factor compare model $m = 1$ with model $m = 2$. The significance boundaries as indicated apply only for the comparison of two models. As the posterior model probability P($m$|$\mathbf{y}$) allows for simultaneous selection among any number of models, the size of the model space has to be taken into account for interpretation

$$P(m|\mathbf{y}) = \frac{p(\mathbf{y}|m)}{\sum\limits_{\mu\in\mathcal{M}} p(\mathbf{y}|\mu)} = \frac{e^{F_m}}{\sum\limits_{\mu\in\mathcal{M}} e^{F_\mu}}. \tag{2.55}$$

The posterior model probability is normalized to the model space $\mathcal{M} = \{1, ..., M\}$ with $\sum_{m\in\mathcal{M}} P(m|\mathbf{y}) = 1$ and is interpreted as the probability that model $m$ is the best model, given the observed data *and* given all evaluated models. Table 2.1 gives an overview for the interpretation of Bayes factors BF, log-Bayes factors log(BF), and posterior model probabilities P($m$|$\mathbf{y}$). Note that the indicated significance boundaries apply for the comparison of two models only. To calculate P($m$|$\mathbf{y}$) for a model space $\mathcal{M}$ with $M > 2$ is feasible, but the interpretation of the achieved posterior model probabilities has to account for the size of the model space. For $M = 3$ models, P($m$|$\mathbf{y}$)$\approx 0.33$ states equal posterior model probabilities.

### 2.4.4  Group Studies

Studies typically contain data from more than one subject. While Mars et al. (2008) and Kolossa et al. (2013) model all subjects at once using third-order GLMs as described in Sect. 2.3.3, there is an emerging tendency in neuroimaging to use subject-specific model estimation with subsequent group-level selection (see, e.g., Garrido et al. 2009; Ostwald et al. 2012; Lieder et al. 2013; Kolossa et al. 2015).

This strict separation of single-subject estimation and group-level selection enables the specification of assumptions about the model distribution over subjects. Specifically, fixed-effects (Stephan et al. 2007) and random-effects analyses (Stephan et al. 2009) can be applied. Penny et al. (2010) give an overview on both approaches: Fixed-effects analyses assume that all subjects use the same model, and should be applied for models of basic functions which are not expected to differ across subjects. Random-effects analyses allow for individual subjects to use different models, and should be used for cognitive tasks which can be solved with different learned strategies. An important drawback of both approaches is the so-called brittleness:

For fixed-effects analyses, the results can become ambiguous if different subjects use different models, or if the model space consists of a large number of models, while for random-effects analyses the addition of just one model to the model space may alter the mutual ranking of all other models (Penny et al. 2010).

Probabilistic inference is assumed to be a basic function and not a consciously available learned scheme. Throughout this work the model space remains small with $M < 20$. Based on these considerations, fixed-effects analyses for group studies will now be introduced and applied in this work.

Given the subject-specific model likelihoods $p(y_\ell|m)$, the group-level likelihood $p(y|m)$ is obtained following (Stephan et al. 2007)

$$p(y|m) = \prod_{\ell=1}^{L} p(y_\ell|m), \tag{2.56}$$

which can be equivalently expressed in log-likelihood and variational free energy as

$$\log(p(y|m)) = \sum_{\ell=1}^{L} \log(p(y_\ell|m)) = F_m = \sum_{\ell=1}^{L} F_{m,\ell}. \tag{2.57}$$

Based on the group log-likelihoods, the group log-Bayes factor log (GBF) (2.53) between two models can be calculated according to

$$\log(GBF_{1\to2}) = \log\left(\frac{p(y|m=1)}{p(y|m=2)}\right) = F_{m=1} - F_{m=2}, \tag{2.58}$$

while the posterior model probability $P(m|y)$ (2.55) for any number of models follows

$$P(m|y) = \frac{p(y|m)}{\sum_{\mu \in \mathcal{M}} p(y|\mu)} = \frac{e^{F_m}}{\sum_{\mu \in \mathcal{M}} e^{F_\mu}}. \tag{2.59}$$

Table 2.1 can be used for the interpretation of group (log-)Bayes factors and posterior model probabilities.

## 2.5 A Transfer Example Experiment—Setup

This section introduces an exemplary experiment which illustrates parameter estimation and model selection as employed in this work. The experiment transfers the methods to the field of speech communication in order to show how widely applicable they are, but the origins of event-related potentials will not be left completely out of sight. This example provided the initial idea for (Kolossa et al. 2016).

Any proposed speech communication system has to be evaluated with regard to the degree to which it distorts the clean speech signal and how these distortions are perceived by humans. Mean opinion scores of listening quality subjective (MOS-LQS) are obtained in formal listening tests, where test subjects are presented with speech samples, the quality of which they rate from 1 (very bad) to 5 (excellent) (ITU 2006a, 2006b). These tests are important, as human perception is the absolute reference concerning speech signal quality, but they are also expensive and time consuming. ITU-T Recommendation Q.862 (ITU 2007) defines the perceptual evaluation of speech quality (PESQ) for automatized and fast speech quality assessment. These mean opinion scores of listening quality objective (MOS-LQO) range from 1 (very bad) to 4.5 (excellent).

The comparison of MOS-LQS with MOS-LQO values is not simple and straightforward. The sequences of MOS-LQS and MOS-LQO values can be interpreted as signals, and while at first glance MOS-LQS seems like a clean signal which captures the listening quality ratings of the subjects, it is in fact corrupted by noise. This noise consists of inter-subject as well as intra-subject contributions (Holmes and Friston 1998), the same as for neuroimaging data in general (Friston et al. 2002) and a sequence of ERP amplitudes in particular (Mars et al. 2008). Inter-subject noise captures subject-individual differences in speech quality perception, which are, for example, different score offsets or individual sensitivity to speech distortions. Intra-subject noise subsumes systematic score deviations which are, e.g., a dependency on the order the files were listened to, and random deviations which occur due to non-deterministic behavior of humans: the same person might rate the same file differently on multiple occasions (Carron and Bailey 1969). Listening tests are designed to reduce this noise by presenting the same file multiple times with different adjacent files and then average the results, but the noise is not eliminated completely. If an alternative model of MOS-LQO is challenging PESQ as the best approximation for the MOS-LQS values, noise has to be taken into account during the model selection phase. Parametric empirical Bayes (see Sect. 2.4) offers a useful tool for evaluating models of speech quality.

In this example experiment, PESQ is assumed to be the ground truth model for speech quality perception. Two kinds of noise are added with different signal-to-noise ratios (SNR) to simulate inter- and intra-subject variability yielding synthetic noisy MOS-LQS values. The MOS-LQO values obtained from the ground truth model, a model which is similar to but distinct from the ground truth model, the null model and the encompassing model are used as competing models, which enter model estimation and selection using parametric empirical Bayes (see Sect. 2.4), log-Bayes factors, and posterior model probabilities (see Sects. 2.4.3 and 2.4.4). It will be shown how confidently the methods used in this work identify the correct model in dependence on the SNR, and how the results depend on the number of trials $N$, as these are the marginal conditions for model selection (Penny 2012).

The rest of this section is structured as follows: First, an overview on the computation of the signal-to-noise ratios is given, followed by a description of the data-generating framework and test conditions. Next, the four competing models entering model selection are introduced. The selection is done for a single subject and a group of 16 subjects to show the effect of both intra- and inter-subject variability. A summary of the evaluation results concludes this chapter.

### 2.5.1  Signal-to-Noise Ratio Simulation

The signal-to-noise ratio simulation for this tutorial is based on the same signal model (1.6) as used in Sect. 1.2 but with a time-variant clean signal $s(n)$. Modeling $s(n)$ to be time-variant over trials is realistic for ERP sequences (Squires et al. 1976; Mars et al. 2008; Kolossa et al. 2013, 2015), which makes the results of these simulations well applicable to SNRs obtained from real ERP sequences with the methods proposed in Sect. 1.2. To recapitulate, the measured signal $y(n)$ is modeled to be composed of a clean signal $s(n)$ and the additive noise $\epsilon(n)$:

$$y(n) = s(n) + \epsilon(n), \tag{2.60}$$

with $n \in \{1, ..., N\}$. Note that in contrast to (1.6) no stimulus-specific index is necessary. This signal model is similar to the one-level linear model (2.11), but with a known clean signal $s(n)$ instead of the estimate $\widehat{s}(n)$ and the term "noise" instead of "error". As in (1.7) the signal-to-noise power ratio (SNR) is defined as

$$\text{SNR} = \frac{P_s}{P_\epsilon}, \tag{2.61}$$

which is in decibel

$$\text{SNR [dB]} = 10 \log_{10} \left( \frac{P_s}{P_\epsilon} \right), \tag{2.62}$$

with the signal power

$$P_s = \frac{1}{N} \sum_{n=1}^{N} s^2(n) \tag{2.63}$$

and the noise power

$$P_\epsilon = \sigma_\epsilon^2. \tag{2.64}$$

Inserting (2.64) in (2.62) and solving for $\sigma_\epsilon^2$ yields

$$\sigma_\epsilon^2 = \frac{P_s}{10^{\frac{\text{SNR [dB]}}{10}}}, \tag{2.65}$$

which can be used to calculate the necessary variance $\sigma_\epsilon^2$ of the error for a desired SNR [dB], given a signal $s(n)$.

### 2.5.2  Synthetic Data and Experimental Conditions

This section describes the framework for the generation of the synthetic data and
test conditions. A note on the version of PESQ which is employed in this section:
When calculating MOS-LQO values PESQ integrates speech signal disturbances
over frequency and time into two factors which capture symmetric and asymmetric
disturbances (Rix et al. 2001). In order to give here a example that is easily trans-
ferable to ERP data, only the symmetric disturbance is included in the PESQ model.
Furthermore, the disturbances are not derived by using clean and degraded speech
signals but are simply sampled, which has no influence on the methods and results but
enables a clear and an easy-to-follow presentation. In order to transfer this example
to model estimation and selection for event-related potentials, the ERP amplitudes
take the place of the MOS-LQO values, while the observer models and link functions
yield the predictors.

The employed simplified version of PESQ follows (Rix et al. 2001)

$$\text{PESQ}(n) = 4.5 - d_{\text{sym}}(n) \cdot 0.1, \tag{2.66}$$

with $d_{\text{sym}}(n) \in \{0, ..., 35\}$ as the symmetric disturbance of speech sample $n$. The
specific formula (2.66) can be generalized to a linear model with two parameters

$$s(n) = \theta_1 - d_{\text{sym}}(n)\theta_2. \tag{2.67}$$

These parameters are used to introduce inter-subject variability, with $\theta_1 \sim \mathcal{N}(4.5, \sigma_{\theta_1}^2)$
and $\theta_2 \sim \mathcal{N}(0.1, \sigma_{\theta_2}^2)$ as random Gaussian variables with means 4.5 and 0.1, and
variances $\sigma_{\theta_1}^2$ and $\sigma_{\theta_2}^2$, respectively. While this is no noise in the classical sense, the
same definitions as in Sect. 2.5.1 are used to make the magnitude of randomness vivid.
Adding zero-mean Gaussian noise $\epsilon \sim \mathcal{N}(0, \sigma_\epsilon^2)$ to (2.67) models the intra-subject
variability

$$y(n) = s(n) + \epsilon(n) = \theta_1 - d_{\text{sym}}(n)\theta_2 + \epsilon(n). \tag{2.68}$$

The integer values for dynamic distortions $d_{\text{sym}}(n)$ are sampled from a uniform distribution in the interval [0, 35], and (2.63) along with (2.65) are used to calculate the
variance $\sigma_\epsilon^2$ necessary to create noise conditions of SNR [dB] $\in \{8, 6, 4, 2, 0, -2\}$.
While these SNR values are not expected for intra-subject variability in speech
quality perception, they are realistic for EEG data (see Sects. 3.6.1 and 4.7.1 for
subject-specific SNR values obtained in the studies in this work). The number of
trials varies with $N \in \{50, 100, 150, 200, 250, 300, 350, 400, 450, 500\}$, yielding 60
scenarios with different combinations of SNR and trial numbers. For the simulation
with multiple subjects, the scenarios are identical for each subject. After subject-
individual model estimation, fixed-effects analyses are applied in order to get the
group-level results (see Sect. 2.4.4). The estimation for all scenarios is repeated a
thousand times with new error and stimulus sampling. (Group) log-Bayes factors
and posterior model probabilities are calculated for each repetition and subsequently

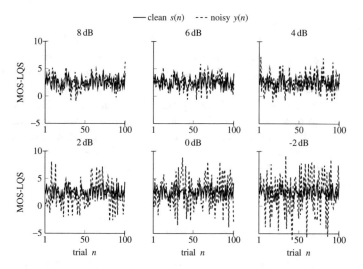

**Fig. 2.1** Clean (——) and noisy (- - -) synthetic MOS-LQS values over trials $n \in \{1, ..., N = 100\}$ for decreasing SNR [dB] $\in \{8, 6, 4, 2, 0, -2\}$

averaged over repetitions (Penny 2012). Figure 2.1 shows clean (——) and noisy (- - -) synthetic MOS-LQS values with parameters $\theta_1 = 4.5$ and $\theta_2 = 0.1$ in the scenarios with $N = 100$ trials for a single repetition and all SNR conditions. The clean signal becomes gradually more disturbed and is unrecognizable for the lowest SNRs.

### 2.5.2.1 A Single Subject

For a single subject the parameters are set to the distribution means ($\theta_1 = 4.5$ and $\theta_2 = 0.1$), and the noisy signal is calculated according to (2.68). The results are presented later on in Sect. 2.6.1 as log-Bayes factors (log(BF), (2.53)) and posterior model probabilities ($P(m|\mathbf{y})$, (2.55)), both based on the variational free energy $F_m$ (2.50).

### 2.5.2.2 Multiple Subjects

For $L = 16$ subjects inter-subject variability is introduced by sampling the parameters $\theta_{\ell,1}$ and $\theta_{\ell,2}$, $\ell \in \mathscr{L} = \{1, 2, ..., L = 16\}$, from their respective distributions $\theta_1 \sim \mathcal{N}(4.5, \sigma_{\theta_1}^2)$ and $\theta_2 \sim \mathcal{N}(0.1, \sigma_{\theta_2}^2)$. Standard deviations are set to $\sigma_{\theta_1} = 0.45$ and $\sigma_{\theta_2} = 0.01$, i.e., the inter-subject parameter variability corresponds to an SNR of 20 dB (2.62). The variational free energy $F_{m,\ell}$ is calculated for each subject within one scenario and then summed over the subjects to obtain $F_m$ (2.57), which is used to calculate group log-Bayes factors (log(GBF), (2.58)) and posterior model probabilities ($P(m|\mathbf{y})$, (2.59)) for model selection in Sect. 2.6.2.

### 2.5.3 The Model Space

This section introduces the four models which make up the model space $\mathcal{M} =$ {TRU, SQD, NUL, ENP} and specifies the respective design matrices which are input to the PEB estimation framework (see Sect. 2.4). The ground truth (TRU) model is the data-generating model, while the squared distortion (SQD) model assumes a quadratic instead of a linear dependence on the symmetric distortion $d_{sym}(n)$. The null (NUL) model is the least complex hypothesis, claiming that all variability in the data is due to noise and will be used as reference model for the log-Bayes factors. The encompassing (ENP) model fits the data perfectly, but is the most complex conceivable model.

#### 2.5.3.1 The Ground Truth Model (TRU)

The ground truth (TRU) model correctly assumes the estimated signal $\widehat{s}(n)$ to depend on an offset $\theta_1$ minus the dynamic distortion $d_{sym}(n)$ multiplied with $\theta_2$:

$$\widehat{s}(n) = \theta_1 - d_{sym}(n)\theta_2, \tag{2.69}$$

which is a two-parameter linear model (2.10). Consequently, the first-level design matrix is of the form

$$\mathbf{X}^{(1)} = \begin{bmatrix} 1 & -d_{sym}(n=1) \\ \vdots & \vdots \\ 1 & -d_{sym}(n=N) \end{bmatrix} \in \mathbb{R}^{N \times 2}. \tag{2.70}$$

#### 2.5.3.2 The Squared Distortion Model (SQD)

The squared distortion (SQD) model states that $\widehat{s}(n)$ depends on an offset $\theta_1$ minus the square of the dynamic distortion

$$\widehat{s}(n) = \theta_1 - d_{sym}^2(n)\theta_2, \tag{2.71}$$

yielding the first-level design matrix

$$\mathbf{X}^{(1)} = \begin{bmatrix} 1 & -d_{sym}^2(n=1) \\ \vdots & \vdots \\ 1 & -d_{sym}^2(n=N) \end{bmatrix} \in \mathbb{R}^{N \times 2}. \tag{2.72}$$

### 2.5.3.3 The Encompassing Model (ENP)

The encompassing (ENP) model assumes that all trials are independent of each other without any additive noise

$$y(n) = \widehat{s}(n) = \theta_n. \tag{2.73}$$

The corresponding first-level design matrix is an identity matrix

$$\mathbf{X}^{(1)} = \mathbf{I}_N = \begin{bmatrix} 1 & 0 & \cdots & 0 \\ 0 & 1 & \ddots & \vdots \\ \vdots & \ddots & \ddots & 0 \\ 0 & \cdots & 0 & 1 \end{bmatrix} \in \mathbb{R}^{N \times N}, \tag{2.74}$$

which models each trial with an independent predictor. Note that the data can be modeled perfectly, but at the cost of highest complexity.

### 2.5.3.4 The Null Model (NUL)

The null (NUL) model represents the least complex hypothesis, claiming that all variation in the data is due to noise. It models $\widehat{s}(n)$ as constant over trials

$$\widehat{s}(n) = \theta_1. \tag{2.75}$$

The first-level design matrix is an all-one column vector

$$\mathbf{X}^{(1)} = \mathbf{x}^{(1)} = \begin{bmatrix} 1 \\ \vdots \\ 1 \end{bmatrix} \in \mathbb{R}^N. \tag{2.76}$$

The NUL model is used as common reference model for the (group) log-Bayes factors, which is inspired by classical null hypothesis testing (see Sect. 2.2.1).

## 2.6 A Transfer Example Experiment—Results

This section shows the results of Bayesian model estimation and selection depending on the number of trials $N$ and the signal-to-noise ratio. It starts with Fig. 2.2 which shows the clean (——) and fitted (- - -) MOS-LQS values for the TRU model for the scenarios depicted in Fig. 2.1. The means of the optimized parameter densities $\mu_{\theta_1|y} = \theta_1$ and $\mu_{\theta_2|y} = \theta_2$ are written in each panel and used as point estimates for the model fitting (see Sect. 2.4.2). The difference between the clean

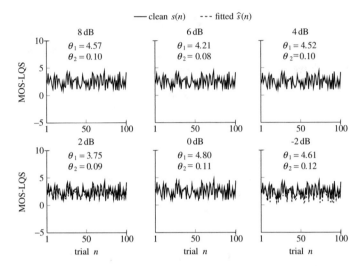

**Fig. 2.2** Clean (——) and fitted (- - -) MOS-LQS values for trials $n \in \{1, ..., N = 100\}$ and SNR [dB] $\in \{8, 6, 4, 2, 0, -2\}$. The fitted MOS-LQS values are based on the TRU model and the point estimates of the parameters $\theta_1$ and $\theta_2$, which are shown in the plots above the respective curves

and fitted MOS-LQS values is graphically indeterminable for SNRs higher than 4 dB. The fit between clean and fitted MOS-LQS values is still good for lower SNRs, but errors become visible.

The rest of this section is split between Bayesian model selection based on a single subject and a group of $L = 16$ subjects. The respective selection results start with Figs. 2.3 and 2.5 depicting the (group) log-Bayes factors ((2.53) and (2.58)) followed by Figs. 2.4 and 2.6 showing posterior model probabilities ((2.55), (2.59)).

## 2.6.1 A Single Subject

Figure 2.3 shows the log-Bayes factors log(BF) of the TRU (—✳—), SQD (- ⊖ -), and ENP ( ⋯ ▫ ⋯ ) models versus the NUL model over an increasing number of trials $N$ in noise conditions SNR [dB] $\in \{8, 6, 4, 2, 0, -2\}$. Throughout all conditions, an increasing number of trials $N$ is accompanied by nearly linearly growing log-Bayes factors for the TRU and SQD model, while for the ENP model the log-Bayes factor is rapidly decreasing. The log-Bayes factors of the SQD model remain constantly below those the TRU model, meaning that the TRU model is favored over the SQD model for any number of trials and all SNR conditions. For an SNR of 8 dB, the TRU model is superior to all other models regardless of the number of trials. At 6 and 4 dB SNR and $N = 50$ trials, all log-Bayes factors are negative, e.g., the NUL model has the greatest variational free energy, while the TRU model is superior for $N = 100$ and more trials. The number of trials necessary for a positive log-Bayes factor for

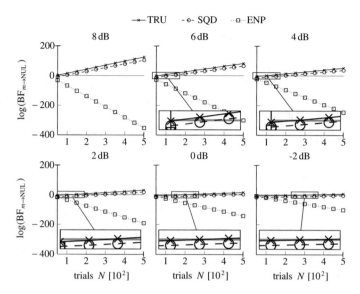

**Fig. 2.3** Log-Bayes factors $\log(\mathrm{BF}_{m\to\mathrm{NUL}})$ (2.53) with $m \in \{\mathrm{TRU},\ \mathrm{SQD},\ \mathrm{ENP}\}$ in dependence on the number of trials $N$ and $\mathrm{SNR}\,[\mathrm{dB}] \in \{8, 6, 4, 2, 0, -2\}$ for a single subject

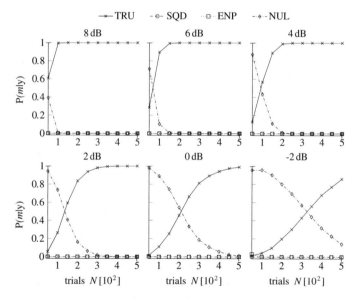

**Fig. 2.4** Posterior model probabilities $P(m|\mathbf{y})$ (2.54) with $m \in \{\mathrm{TRU},\ \mathrm{SQD},\ \mathrm{ENP},\ \mathrm{NUL}\}$ in dependence on the number of trials $N$ and $\mathrm{SNR}\,[\mathrm{dB}] \in \{8, 6, 4, 2, 0, -2\}$ for a single subject

the TRU model versus the NUL model increases with decreasing SNR. For 2 dB, 0 dB, and $-2$ dB, the necessary number of trials is $N = 150$, $N = 250$, and $N = 350$, respectively. The area at which the NUL model ceases to be supported is magnified in the lower parts of the panels.

Posterior model probabilities $P(m|\mathbf{y})$ (2.55) offer a more intuitive way of model selection. Figure 2.4 shows posterior model probabilities for the TRU ($-\!\!*\!\!-$), SQD (-⊖-), ENP (⋯□⋯), and NUL (-⋄-) models in dependence on the number of trials $N$ in noise conditions SNR [dB] $\in \{8, 6, 4, 2, 0, -2\}$. Naturally, they mirror the results obtained from Fig. 2.3, as they are also based on the variational free energy, but the characteristic areas where the model with the highest posterior model probability changes are much clearer. In contrast to the steady increase in log-Bayes factors, posterior model probabilities change more rapidly around the critical trial number necessary for correctly identifying the TRU model as the best model. For SNRs of 0 and $-2$ dB, the statistical significance even at high trial numbers of $N = 250$ and $N = 350$ is still very low with posterior model probabilities of $P(\text{TRU}|\mathbf{y}) \approx 0.65$ and $P(\text{TRU}|\mathbf{y}) \approx 0.55$, respectively. Thus, inference based on a single participant in very noisy conditions is not feasible or requires a lot of trials.

## 2.6.2 Multiple Subjects

Figure 2.5 shows the group log-Bayes factors (log(GBF), i.e., log-Bayes factors summed over subjects (2.58)) of the TRU ($-\!\!*\!\!-$), SQD (-⊖-) and ENP (⋯□⋯) models versus the NUL model over an increasing number of trials $N$ in noise conditions SNR [dB] $\in \{8, 6, 4, 2, 0, -2\}$. The results are qualitatively very similar to those obtained for a single subject in Fig. 2.3, i.e., linear evolution of group log-Bayes factors for an increasing number of trials $N$ in all noise conditions, but the absolute values are on a larger scale. For conditions with an SNR of 6 dB or lower and only a few trials, the log(GBF) favors the NUL model. The number of trials necessary for correctly identifying the TRU model as the best model and rejecting the NUL are close to those for a single subject, i.e., $N = 100$, $N = 100$, $N = 150$, $N = 200$, and $N = 300$ for 6 dB, 4 dB, 2 dB, 0 dB, and $-2$ dB, respectively.

Figure 2.6 shows posterior model probabilities (2.59) for the TRU ($-\!\!*\!\!-$), SQD (-⊖-), ENP (⋯□⋯), and NUL (-⋄-) models in dependence on the number of trials $N$ for noise conditions SNR [dB] $\in \{8, 6, 4, 2, 0, -2\}$. As expected, the results mirror those depicted in Fig. 2.5, while meaningful differences are revealed in comparison to Fig. 2.4: Up to 2 dB SNR the critical trial numbers are identical to those of a single subject, but for a signal-to-noise ratio of 0 and $-2$ dB, $N = 200$ and $N = 300$ trials suffice for correctly identifying the TRU model as best model. In each case these are 50 trials less than for a single participant, for whom the posterior model probabilities of about 0.5 did not permit to draw meaningful conclusions at these points. Over all data points the statistical significance is greatly increased in comparison to Fig. 2.4 allowing for much more reliable conclusions (see Table 2.1 for details).

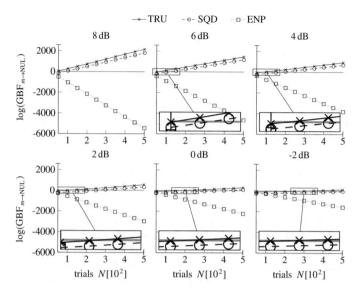

**Fig. 2.5** Group log-Bayes factors $\log(\text{GBF}_{m\to\text{NUL}})$ (2.58) with $m \in \{\text{TRU, SQD, ENP}\}$ in dependence on the number of trials $N$ and for noise conditions $\text{SNR}\,[\text{dB}] \in \{8, 6, 4, 2, 0, -2\}$ for a group of $L = 16$ subjects

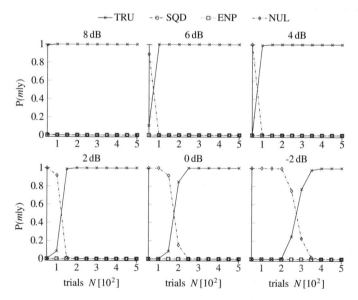

**Fig. 2.6** Posterior model probabilities $\text{P}(m|\mathbf{y})$ (2.59) with $m \in \{\text{TRU, SQD, ENP, NUL}\}$ in dependence on the number of trials $N$ and for noise conditions $\text{SNR}\,[\text{dB}] \in \{8, 6, 4, 2, 0, -2\}$ for a group of $L = 16$ subjects

## 2.7  Evaluation Summary

Log-Bayes factors and posterior model probabilities derived in the parametric empirical Bayes (PEB) framework are proven to be very useful for model estimation and selection. The use of this method is limited by the quality and quantity of the data, i.e., the signal-to-noise ratio (SNR) and the number of trials $N$. In view of the variability of SNRs across subjects, $N \geq 200$ trials per subject should be sufficient for reliably identifying the best model (see Sects. 3.6.1 and 4.7.1 for subject-specific $\widehat{\text{SNR}}$ values obtained in this work).

A major shortcoming of log-Bayes factors is that only two models are directly compared with each other simultaneously, which makes it complicated to get a clear view of the results for the complete model space. Additionally, (group) log-Bayes factors can become extremely large (see Fig. 2.5 with group log-Bayes factors in the magnitude of $10^3$), which leads to a tendency to judge small (group) log-Bayes factors as not appropriately significant (see Table 2.1 with log-Bayes factors greater than five denoting very strong significance). Posterior model probabilities do not suffer from these shortcomings. All models composing the model space are simultaneously compared to each other, and the probabilities are normalized to the model space. The interpretation of statistical significance is intuitive and reminiscent of classical approaches, making Bayesian model selection also accessible to non-experts. Taking multiple subjects into account decreases the required number of trials per subject for correctly selecting the TRU model under low signal-to-noise ratios and increases the statistical power of the results. Group studies are therefore mandatory in order to report meaningful results.

# Chapter 3
# A New Theory of Trial-by-Trial P300 Amplitude Fluctuations

This chapter reviews state-of-the-art observer models of the P300 event-related potential and introduces a new digital filtering (DIF) model. It starts with a brief overview of the models known from literature and of the approach proposed in this work. After a description of the employed variant of the oddball task and specific methods for capturing trial-by-trial fluctuations in the P300 amplitudes $y(n)$, the models and response functions constituting the model space $\mathcal{M}$ are presented in detail and the two most renowned ones are integrated into the digital filtering model. Next, the parameter optimization schemes as well as the composition of the design matrices for model estimation and selection (see Chap. 2) are specified. Results and conclusions complete this chapter, which was adapted and extended from Kolossa et al. (2013).

## 3.1 Overview

In the following, the oddball task and the four observer models investigated in this chapter are briefly described. The subjects performed an oddball task, which is a simple two-choice response time task: Frequent and rare events are sequentially shown to the subjects who have to respond to the event type by pressing the corresponding key on a keyboard. The trial-by-trial P300 amplitudes were derived from the EEG signal at electrode Pz and used for Bayesian model estimation and selection (see Sect. 2 for details).

Table 3.1 presents all observer models: The first model is called here SQU and goes back to Squires et al. (1976), who showed and modeled dependencies of P300 amplitude fluctuations from observable events based on the concept of expectancy, which was thought to be determined by three factors: "(i) the memory for event frequency within the prior stimulus sequence, (ii) the specific structure of the prior sequence, and (iii) the global probability of the event." (p. 1144). Note that as expectancy is not a probability, it is not linked to the P300 amplitudes via a response function but used to predict the amplitudes directly. Although Squires et al.'s model offers

© Springer International Publishing Switzerland 2016
A. Kolossa, *Computational Modeling of Neural Activities for Statistical Inference*, DOI 10.1007/978-3-319-32285-8_3

**Table 3.1** Overview and short description of the observer models

| Model | Description |
|-------|-------------|
| SQU | Observer model based on memory with exponential forgetting, alternation expectation, and prior knowledge (Squires et al. 1976) |
| MAR | Observer model based on memory with no forgetting (Mars et al. 2008) |
| OST | Observer model based on memory with exponential forgetting (Mars et al. 2008) |
| DIF | Observer model which fuses properties of the SQU and MAR models using a digital filtering approach (Kolossa et al. 2013) |

a good explanation of the measured data, it remained descriptive and not plausible. Specifically, Squires et al. (1976) incorporated a lookup table instead of a formal calculus for alternation expectation as well as knowledge which was not available to the subjects. In order to make the SQU model accessible to model selection, it will be reformulated in a completely computational way in Sect. 3.3.1.

The second investigated model is called MAR and goes back to Mars et al. (2008), who proposed a computational model of processes underlying the generation of the P300 amplitudes in which trial-by-trial amplitude fluctuations are explained by an observer keeping track of the global probability distribution over events. This distribution is linked to the P300 amplitudes via predictive surprise (see Sect. 1.4.2), but postdictive surprise (see Sect. 1.4.1) is alternatively tested as a response function in this work. The subjective estimates of statistical regularities in the environment are modeled to depend solely on the integration of observations over infinitely long periods of time. This is similar to factor (iii) of the SQU model, but in contrast to the SQU model the statistical regularities are learned from observations with uniform initial prior probabilities and are not assumed to be known a priori. However, the explanatory power of this observer model is limited, because it cannot account for the effects of the recent stimulus sequence on the P300 amplitudes, which were already well-documented decades earlier (e.g., Squires et al. 1976; Leuthold and Sommer 1993).

The third investigated model is called OST in this work and goes back to Ostwald et al. (2012), who proposed a computational model that treats the event probability itself as a hidden random variable. They used Bayesian surprise as response function (see Sect. 1.4.1) to link the trial-by-trial evolution of the distribution over the hidden variable to neural activities. Their model is based on exponentially forgetting memory traces similar to factor (i) of the SQU model, but with a uniform initial prior such as the MAR model. But as the MAR model, this model cannot account appropriately for the influence of the structure of the recent stimulus sequence on P300 amplitude fluctuations.

This chapter also introduces a computational digital filtering (DIF) model which integrates short-term, long-term, and alternation-based contributions to calculate the probability distribution over events. Specifically, subjective estimates of statistical regularities in the environment are kept at short-term and long-term decay time parameters (similar to factors (i) and (iii) in Squires et al. 1976). It implements an

**Fig. 3.1** Hierarchical structure that relates the observable random variables and the DIF model to neural activities. Parts of the full structure which are not covered by the oddball task and the DIF model are outlined in *light gray*. The DIF model updates the probability distribution over observable events $k$. Postdictive surprise and predictive surprise are the response functions that link the coding and computing of the probability distribution to neural activities

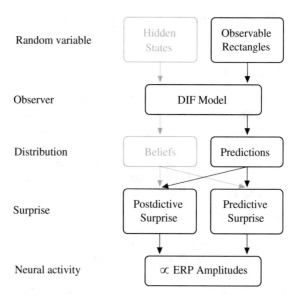

alternation term (similar to factor (ii) in Squires et al. 1976) as well as uniform initial prior probabilities (as Mars et al. 2008). It is further shown that this model can be described as two parallel first-order infinite impulse response (IIR) low-pass filters and an additional fourth-order finite impulse response (FIR) high-pass filter. Postdictive surprise and predictive surprise (see Sect. 1.4) are used as response functions to relate the trial-by-trial changes of the probability distribution to P300 amplitude fluctuations.

Figure 3.1 gives an overview on the part of the hierarchical structure that relates random variables to neural activity, which is covered by the oddball task and the DIF model (see Chap. 1.4 for a formal introduction to the framework of probabilities and surprise). The oddball task consists solely of observable events $k$, which are red and blue rectangles, respectively. The DIF model functions as an observer who estimates predictions over future observations in the form of a probability distribution over observable events. The coding and computing of this probability distribution are linked to the ERP amplitudes via predictive surprise and postdictive surprise as response functions, respectively. There are no hidden random variables in the oddball task and the observer model does not calculate any beliefs. Thus, these parts of the full structure are outlined in light gray. The model space in this chapter consists of $\mathcal{M} = \{\text{SQU}, I_B(\text{MAR}), I_P(\text{MAR}), I_B(\text{OST}), I_B(\text{DIF}), I_P(\text{DIF}), \text{NUL}\}$.

## 3.2  Participants, Experimental Design, Data Acquisition, and Data Analysis

### Participants

Sixteen healthy subjects (fourteen women, two men, mean age of 20 years; age range 18–23 years; mean handedness (Oldfield 1971): 74; handedness range $-76 \dots 100$), all with normal or corrected-to-normal visual acuity participated in the experiment. All were recruited from introductory courses at the Department of Psychology at Technische Universität Braunschweig for course credits. Experimental procedures were approved by the local ethics committee and in accordance with the Declaration of Helsinki.

### Experimental Design

Subjects performed a simple two-choice response time task without feedback about response accuracy in which all stimuli had equal behavioral relevance. This feature of the experimental design constitutes an important difference between this variant and the classical oddball task in which subjects usually discriminate between task-relevant (target) and irrelevant (standard) stimuli (Ritter and Vaughan 1969). There were $K = 2$ types of events $k \in \mathcal{K} = \{1, 2\}$, which formed the visual stimuli presented to the subjects over trials. The stimulus was either a red ($k = 1$) or a blue rectangle ($k = 2$), which was perceived by the subjects taking the form of observation $o(n) = k$ at trial $n$. Subjects were instructed to respond to each stimulus with the previously associated button as quickly as possible, but not at the expense of accuracy. They used the index finger of both hands (e.g., left button on response to the red rectangle, right button in response to the blue rectangle). Stimulus-response mapping (i.e., [red-left, blue-right] or [red-right, blue-left], respectively) was counterbalanced over the subjects.

   The subjects performed $B = 6$ blocks $b \in \{1, \dots, 6\}$ of $N = 192$ trials $n \in \{1, \dots, 192\}$ in two experimental conditions $c \in \mathcal{C} = \{[0.5, 0.5], [0.3, 0.7]\}$. The probability of the occurrence of each event $P(o = k)$ was manipulated between blocks such that the relative probabilities of events were either 0.5 for each event (i.e., $P(o=1) = P(o=2) = 0.5$) across six consecutive blocks (1152 trials overall), or 0.7 for the frequent event $k = 1$ (i.e., $P(o=1) = 0.7$), and 0.3 for the rare event $k = 2$ (i.e., $P(o=2) = 0.3$) across the remaining six consecutive blocks (1152 trials overall). Stimulus-probability mapping was counterbalanced over subjects (i.e., a stimulus color signified the frequent event in 50 % of the subjects but the rare event in the remaining subjects), but this will be ignored in the description to avoid confusion. Throughout this chapter (●) represents the frequent event $k = 1$ and (●) the rare event $k = 2$.

   The order of the probability manipulation was counterbalanced over subjects (experimental condition $c = [0.5, 0.5]$ prior to $c = [0.3, 0.7]$ or vice versa) *who were not informed about these probabilities*. The subjects were informed that the two different stimuli were randomly distributed across blocks. Between the blocks a break was scheduled and the subjects were free to initiate the subsequent block at their own pace.

A single stimulus was presented at each trial for a duration of 100 ms each, with a stimulus to stimulus interval of 1.5 s. Stimuli were displayed at the center of the monitor against a light gray background with a viewing distance of 1.25 m.

**Selection of ERP Data for Further Analyses**

The region of interest (ROI) contained solely the electrode $e = Pz$, as the P300 is traditionally reported to be maximal there (Duncan-Johnson and Donchin 1977; Polich 2007; Mars et al. 2008). A high-pass filter (0.75 Hz cutoff frequency, 48 db/oct) was applied to the digital EEG signal in order to eliminate low-frequency variations, which were associated with the occasional occurrence of electrodermal artifacts. The EEG was then divided into epochs of 1000 ms duration, starting 100 ms before stimulus onset (see (1.1) in Sect. 1.1 for details). Epochs were corrected using the interval $[-100, 0]$ ms before stimulus presentation as the baseline (1.2). In order to find the ERP component latency $t_{ERP}$, event-related potential (ERP) waveforms $\bar{y}_{\ell,k}(t)$ (1.3) were created for each subject $\ell \in \mathscr{L} = \{1, \ldots, L = 16\}$ and for each event type $k \in \{1, 2\}$ in the $[0.3, 0.7]$ experimental condition. The maximum of the difference between the two waves (as in (1.4)) within the whole epoch length determined the ERP component latency $t_{\ell, P300}$. The thus derived $t_{\ell, P300}$ values ($\mu_{t_{ERP}} = 344$ ms, $\sigma_{t_{ERP}} = 48$ ms; $\min_{t_{ERP}} = 280$ ms; $\max_{t_{ERP}} = 464$ ms) were used for the P300 amplitude acquisition in the $[0.5, 0.5]$ experimental condition as well. The P300 amplitudes $y_{\ell,c,b}(n)$ were finally derived following (1.5) with the smoothing interval of $t_{\ell, P300} \pm 60$ ms as in (Mars et al. 2008).

## 3.3 State-of-the-Art Observer Models and Surprise

This section formalizes the state-of-the-art observer models and their relation to surprise (see Sect. 1.4 for an overview on probabilities and surprise). In the following a rigorous mathematical formulation of the approach by Squires et al. (1976) is given, followed by short recapitulations of the approaches by Mars et al. (2008) and Ostwald et al. (2012). Note that all models play the role of a dynamically updated *observer* who learns the statistical parameters of the stimulus sequence. Finally, different kinds of surprise are used as response functions that link the probability distributions to the P300 amplitudes $y(n)$.

### 3.3.1 Approach by Squires et al. (SQU)

Squires et al. (1976) did not formulate a strict computational model to compute the observation probability $P_k(n)$. Instead, having investigated solely a $K = 2$ case, they use the notion of *expectancy* $E_k(n)$ for event $k$ on trial $n$. While Squires et al. (1976) have described their model partly in mathematics, partly in words, a complete analytical formulation of their approach is now presented, which is straightforward to implement in software. Their empirical formulation of expectancy that event $k \in \{1, 2\}$ will be observed on trial $n \in \{1, \ldots, N\}$ is given as

$$E_k(n) = 0.235 \cdot \check{c}_{S,k}(n) + 0.033 \cdot \check{c}_{A,k}(n) + 0.505 \cdot P_k - 0.027, \qquad (3.1)$$

with three expectancy contributions, namely (i) a count function for the *short-term memory*, $\check{c}_{S,k}(n)$, (ii) a count function for the *alternation expectancy*, $\check{c}_{A,k}(n)$, and (iii) the assumed-to-be-known global probability, $P_k$ (and an additive constant). Note that the indices "S" and "A" of the count functions $\check{c}_{S,k}(n)$ and $\check{c}_{A,k}(n)$ express their *short-term memory* and *alternation expectancy* characters, respectively.

(i) **Short-Term Memory**

The short-term memory count function is defined as

$$\check{c}_{S,k}(n) = \sum_{\nu=n-N_{\text{depth}}}^{n-1} \gamma_S^{n-\nu} d_k(\nu), \qquad (3.2)$$

with the time sequence

$$d_k(\nu) = \begin{cases} 1, & \text{if } o(\nu) = k \\ 0, & \text{otherwise} \end{cases} \qquad (3.3)$$

for $\nu = 1, 2, \ldots, n-1$. The count function covers only a limited memory span of $N_{\text{depth}} = 5$ and introduces an exponential forgetting factor $\gamma_S^{n-\nu} = e^{-\frac{n-\nu}{\beta_S}}$ with $0 \leq \gamma_S \leq 1$ and a time constant $0 \leq \beta_S < \infty$, with $\gamma_S = 0.6$ for all experimental conditions (i.e., $\beta_S = 1.96$). Note that the count function (3.2) depends only on event observations in the *recent* past.

(ii) **Alternation Expectancy**

In contrast, the term $\check{c}_{A,k}(n) \in \{-(N_{\text{depth}} - 2), \ldots, -3, -2, 0, +2, +3, \ldots, (N_{\text{depth}} - 2)\}$, denotes the expectancy w.r.t. alternating observations, and how this expectancy is met by the present observation $o(n)$. Squires et al. (1976) use $N_{\text{depth}} = 5$ as for the short-term memory $\check{c}_{S,k}(n)$ (3.2). Following Squires et al. (1976), its sign

$$\varsigma_k(n) = 2|d_k(n) - d_k(n-1)| - 1 \in \{-1, +1\} \qquad (3.4)$$

is negative ($\varsigma_k(n) = -1$) if observation $o(n)$ violates the alternation expectation (i.e., $o(n)$ and $o(n-1)$ are identical). On the other hand, if the alternation expectation is met (i.e., $o(n)$ and $o(n-1)$ differ from each other), the sign is positive ($\varsigma_k(n) = +1$). The number of *previous* alternations *in a row* constitutes the amplitude of the expectation, which is given by

$$N_{\text{alt}} = \arg\max_{N'_{\text{alt}}} \prod_{\nu=1}^{N'_{\text{alt}}} 2\big|d_k(n-\nu) - d_k(n-\nu-1)\big|, \tag{3.5}$$

with $N'_{\text{alt}} \in \{1, 2, \dots, (N_{\text{depth}} - 2)\}$. Finally, $\check{c}_{A,k}(n)$ is given as

$$\check{c}_{A,k}(n) = \varsigma_k(n) \cdot \big(\min\{N_{\text{alt}} - 2, 0\} + N_{\text{alt}}\big). \tag{3.6}$$

The use of the minimum function ensures that an expectation for alternation requires at least $N_{\text{alt}} = 2$ consecutive previous stimulus alternations. As an example, Squires et al. (1976) have given the values of $\check{c}_{A,k}(n)$ for $K = 2$, $N_{\text{depth}} = 5$, and $o(n) = k = 1$ (which shall be denoted as $a$): $\{o(n-4)o(n-3)o(n-2)o(n-1)o(n)\} = bbaba$: $\check{c}_{A,k}(n) = +2$, $ababa$: $\check{c}_{A,k}(n) = +3$, $babaa$: $\check{c}_{A,k}(n) = -3$, $aabaa$: $\check{c}_{A,k}(n) = -2$; all other 12 fourth-order sequences $xxxxa$ terminating with $a$ yield $\check{c}_{A,k}(n) = 0$ according to the minimum function in (3.6).

(iii) **Global Probability**

The term $P_k = P(o(n) = k)$ in (3.1) denotes the true global probability of the observation being event $k$. It is nothing other than the relative frequency of the stimulus in the current experimental condition, which must be made known to this model.

**Summary**

In summary, (3.1) provides a model for expectancy that is linearly composed of three contributions: First, a purely predictive limited length ($N_{\text{depth}} = 5$) exponentially decaying short-term memory (cf. count function $\check{c}_{S,k}(n)$ in (3.2)). Second, an expectancy contribution in the range $-3 \leq \check{c}_{A,k}(n) \leq +3$ depending on the extent to which a first-order alternation ($aa$ or $ba$) expectation has been built up and then met/violated within the latest observed $N_{\text{depth}} = 5$ trials. Third, the relative frequency $P_k$ of event $k$ which equals the correct global probability throughout experimental conditions. Note that the relative frequency $P_k$ is not learned sequentially by experience, but assumed to be known by the subjects.

### 3.3.2  Approach by Mars et al. (MAR)

Mars et al. (2008) propose an observer model without forgetting which keeps track of the observation probability $P_k(n) = P(o(n) = k | \mathbf{o}_1^{n-1})$ according to

$$P_k(n) = \frac{\tilde{c}_{L,k}(n) + 1}{(n-1) + K}, \tag{3.7}$$

where

$$\tilde{c}_{L,k}(n) = \sum_{\nu=1}^{n-1} d_k(\nu) \tag{3.8}$$

counts the number of occurrences[1] of event $k$ until trial $n - 1$. The time sequence $d_k(\nu)$, with $\nu = 1, 2, \ldots, n - 1$, satisfies

$$d_k(\nu) = \begin{cases} 1, & \text{if } o(\nu) = k \\ 0, & \text{otherwise} . \end{cases} \tag{3.9}$$

As can be seen in (3.7), the observation probability for event $k$ on trial $n = 1$ equals a uniform initial prior $P_k(n = 1) = \frac{1}{K}$. After many trials ($n \gg K$, and $\tilde{c}_{L,k}(n) \gg 1$), the observation probability approximates $P_k(n) \approx \frac{\tilde{c}_{L,k}(n)}{n-1}$, i.e., the relative frequency of event $k$ until trial $n - 1$. Note that the index "L" of the count function $\tilde{c}_{L,k}(n)$ (3.8) expresses the *long-term memory* character of the MAR model. After $o(n)$ has been observed on trial $n$, the observation probability $P_k(n + 1) = P(o(n + 1) = k|\mathbf{o}_1^n)$ for trial $n + 1$ can be calculated. Taking $P_k(n)$ and $P_k(n + 1)$ for all $k \in \mathcal{K}$ yields the observation probability distributions $P_{\mathcal{K}}(n)$ and $P_{\mathcal{K}}(n + 1)$, respectively.

### 3.3.3  Approach by Ostwald et al. (OST)

Ostwald et al. (2012) propose a model of surprise for $k \in \{1, 2\}$ observable events with exponential forgetting of past observations. The OST model does not estimate the discrete observation probabilities $P_k(n)$ directly, but treats the observation probability $\chi = P(o(n) = k) \in [0, 1]$ itself as a random variable. The probability density function of a beta distribution over $\chi$ is parameterized via event counters $\tilde{c}_k(n)$:

$$p(\chi|\tilde{c}_1(n), \tilde{c}_2(n)) = \frac{\Gamma(\tilde{c}_1(n) + \tilde{c}_2(n))}{\Gamma(\tilde{c}_1(n))\,\Gamma(\tilde{c}_2(n))} \chi^{\tilde{c}_1(n)}(1 - \chi)^{\tilde{c}_2(n)}, \tag{3.10}$$

with $\Gamma$ being the gamma function. The event counters are updated on a trial-by-trial basis according to

$$\tilde{c}_k(n) = \sum_{\nu=0}^{n} e^{-\frac{1}{\beta}(n-\nu)} \tilde{c}_k'(\nu), \tag{3.11}$$

with $k \in \{1, 2\}$. The parameter $\beta$ controls memory length and

$$\tilde{c}_k'(n) = \sum_{\nu=0}^{n} \tilde{g}_k(\nu), \tag{3.12}$$

---

[1] Note that the number of occurrences of *all* events $k \in \mathcal{K}$ until trial $n-1$ is simply $\sum_{k \in \mathcal{K}} \tilde{c}_{L,k}(n) = n - 1$.

with

$$\tilde{g}_k(n) = \begin{cases} 1, & \text{if } n = 0 \text{ (uniform beta prior distribution)} \\ 1, & \text{if } n > 0 \text{ and } o(n) = k \\ 0, & \text{otherwise,} \end{cases} \qquad (3.13)$$

counts the number of occurrences of event $k$ until trial $n$. The control parameter is set to $\beta = 2.6$ as in the "BS3" model in (Ostwald et al. 2012).

### 3.3.4 Surprise Based on the SQU, MAR, and OST Models

Squires et al. (1976) (see Sect. 3.3.1) assume direct proportionality between the *expectancy* $E_{k=o(n)}(n)$ (3.1) for observation $o(n)$ on trial $n$ to be event $k$ and the trial-by-trial P300 amplitude

$$y(n) \propto E_{k=o(n)}(n). \qquad (3.14)$$

Thus, no surprise is calculated with regard to the SQU model. Mars et al. (2008) (see Sect. 3.3.2) state substantial evidence for *predictive* surprise

$$I_P(n) = -\log_2 P_{k=o(n)}(n) \qquad (3.15)$$

based on the observation probability $P_{k=o(n)}(n)$ (3.7) to be proportional to $y(n)$. In order to consequently apply the surprise framework as introduced in Sect. 1.4, the MAR model is additionally tested with *postdictive* surprise (1.18) based on the observation probability distributions $P_{\mathcal{K}}(n)$ (before observation $o(n)$) and $P_{\mathcal{K}}(n+1)$ (after $o(n)$ has been observed):

$$I_B(n) = D_{KL}\left(P_{\mathcal{K}}(n) \,||\, P_{\mathcal{K}}(n+1)\right) = \sum_{k \in \mathcal{K}} P_k(n) \log\left(\frac{P_k(n)}{P_k(n+1)}\right). \qquad (3.16)$$

Ostwald et al. (2012) (see Sect. 3.3.3) assume $y(n)$ to be proportional to *Bayesian* surprise $I_B(n)$, which reflects the change in the probability density function (3.10) induced by the observation $o(n)$. Consequently, it is calculated as the Kullback–Leibler divergence between the probability density functions $p(\chi|\tilde{c}_1(n-1), \tilde{c}_2(n-1))$ (before observation $o(n)$) and $p(\chi|\tilde{c}_1(n), \tilde{c}_2(n))$ (after $o(n)$ has been observed) (Ostwald et al. 2012)

$$I_B(n) = \int p(\chi|\tilde{c}_1(n-1), \tilde{c}_2(n-1)) \log\left(\frac{p(\chi|\tilde{c}_1(n-1), \tilde{c}_2(n-1))}{p(\chi|\tilde{c}_1(n), \tilde{c}_2(n))}\right) d\chi. \qquad (3.17)$$

Note that as $p(\chi)$ is a probability density function and not a discrete probability, predictive surprise and entropy cannot be calculated based on the OST model.

## 3.4   The Digital Filtering Model (DIF)

This section presents the digital filtering (DIF) model, which keeps track of the observation probability $P_k(n)$ and is inspired by both Squires et al. (1976) and Mars et al. (2008). It was developed with the aim to unify the exponentially decaying short-term memory and alternation expectation capabilities of Squires et al. (1976) with the learned relative frequency estimation of Mars et al. (2008) (long-term), and to express the result in terms of a simple new digital filtering (DIF) model. Besides an additive probability normalizing constant $\frac{1}{C}$, it consists of three additive contributions to the observation probability, each of which can be represented by a digital filter: A long-term contribution (L), a short-term one (S), and one term capturing alternations (A) as depicted in Fig. 3.2:

$$P_k(n) = \alpha_L \cdot c_{L,k}(n) + \alpha_S \cdot c_{S,k}(n) + \alpha_A \cdot \left[ c_{A,k}(n) + \tfrac{1}{C} \right]. \qquad (3.18)$$

There are three different count functions used, each represented by a digital filter transfer function $H(f)$ applied to the common input signal $g_k(n)$, which is given as

$$g_k(n) = \begin{cases} \frac{1}{K}, & \text{if } n \leq 0 \text{ (uniform initial prior)} \\ 1, & \text{if } n > 0 \text{ and } o(n) = k \\ 0, & \text{otherwise.} \end{cases} \qquad (3.19)$$

This signal implicitly contains an initial prior of $\frac{1}{K}$ at the start of a block of trials, and a "1" wherever the observation equals $o(n) = k$, otherwise a "0". As the subjects were not informed about event probabilities, a uniform initial prior $P_k(1) = \frac{1}{K}, \forall k \in \{1, \dots, K\}$ is a reasonable model assumption before any stimulus has been observed. Note that in contrast to the sequence $d_k(\nu)$ ((3.3), (3.9)) as used in Squires et al. (1976) (3.2) and Mars et al. (2008) (3.8), (3.19) defines a model-exciting infinite length *signal* $g_k(\nu)$, $\nu \in \{-\infty, \dots, n-2, n-1\}$. The digital filter model yields an

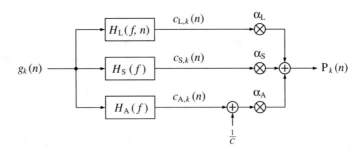

**Fig. 3.2** Block diagram of the new digital filter (DIF) model with input $g_k(n)$ (3.19) and output $P_k(n)$ (3.18), digital filter transfer functions $H(f)$, with $0 \leq f \leq f_{sp}/2$, a stimulus presentation rate of $f_{sp} = 2/3\,\text{Hz}$, and probability normalizing constant $\frac{1}{C}$

output signal $P_k(n)$ as given in (3.18). The weighting parameters $\alpha_L, \alpha_S, \alpha_A$ hold $\alpha_L + \alpha_S + \alpha_A = 1$ and $0 \leq \alpha_{i'} \leq 1$, $i' \in \{L, S, A\}$. Note that calculating (3.18) for all $k \in \mathcal{K}$ yields the observation probability distribution $P_{\mathcal{K}}(n)$.

### 3.4.1 Short-Term Memory

The short-term memory count function can be expressed as

$$c_{S,k}(n) = \frac{1}{C_S} \sum_{\nu=-\infty}^{n-1} \gamma_S^{n-\nu} g_k(\nu), \qquad (3.20)$$

with some normalizing constant $C_S = \frac{\gamma_S}{1-\gamma_S}$ and an exponential forgetting factor $\gamma_S^{n-\nu} = e^{-\frac{n-\nu}{\beta_S}}$ with $0 \leq \beta_S < \infty$, as with count function $\check{c}_{S,k}(n)$ in (3.2). The transfer function of the short-term digital filtering process as described by (3.20) is depicted in Fig. 3.2 as $H_S(f)$, and is plotted in Fig. 3.3 as (- ·--·-), revealing a smooth (i.e., weak) low-pass characteristic. Note that the short-term memory count function (3.20) can be expressed mathematically equivalent in a recursive form according to

$$c_{S,k}(n) = (1 - \gamma_S) \cdot g_k(n-1) + \gamma_S \cdot c_{S,k}(n-1), \qquad (3.21)$$

initialized with the uniform initial prior $c_{S,k}(0) = \frac{1}{K}$, which in (3.20) was contained in the values $g_k(\nu) = \frac{1}{K}$ for $\nu \leq 0$ of (3.19). The recursive character of (3.21) becomes apparent by substituting the right hand side of (3.21) in itself for calculating $c_{S,k}(n-1)$. Figure 3.4 shows the block diagram of the infinite impulse response

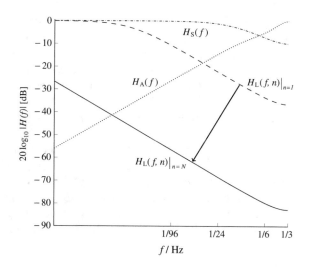

**Fig. 3.3** Amplitude responses of the long-term ($H_L(f, n)$, (- - -) and (——)), short-term ($H_S(f)$, (- ·--·-)), and alternation ($H_A(f)$, (······)) filters of the DIF model as a function of the input signal frequency $f$ (logarithmic scale!). The dynamic long-term filter is shown for $n = 1$ (- - -) and $n = N = 192$ (——)

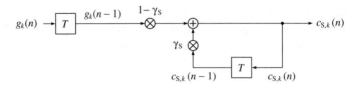

**Fig. 3.4** Block diagram of the first-order infinite impulse response length (IIR) filter with transfer function $H_S(f)$, which captures the short-term memory equivalent to (3.20) or (3.21). Elements $\boxed{T}$ denote a delay of one trial. At the adder element $\oplus$, updating of the weighted output $\gamma_S \cdot c_{S,k}(n-1)$ of the last trial $n-1$ with the weighted input $(1-\gamma_S) \cdot g_k(n-1)$ of the last trial yields the current output $c_{S,k}(n)$. Note that via $\gamma_S \cdot c_{S,k}(n-1)$, all preceding inputs (and outputs) influence the current output, though for the short-term memory the influence of trials in the long past is negligible

(IIR) digital filter which captures the short-term memory and illustrates the updating process inherent in (3.21). The input signal defined in (3.19) and the weights $(1-\gamma_S)$ and $\gamma_S$ guarantee $0 \le c_{S,k}(n) \le 1$. The equivalence of (3.20) and (3.21) and the derivation of $C_S = \frac{\gamma_S}{1-\gamma_S}$ are shown in the Appendix.

### 3.4.2  Long-Term Memory

The long-term memory count function can be expressed as

$$c_{L,k}(n) = \frac{1}{C_{L,n}} \sum_{\nu=-\infty}^{n-1} \gamma_{L,n}(\nu)\, g_k(\nu), \tag{3.22}$$

with the time-variant (i.e., *dynamic*) exponential forgetting factor $\gamma_{L,n}(\nu) = \prod_{\upsilon=\nu+1}^{n}$ $\gamma_{L,\upsilon} \frac{1-\gamma_{L,\upsilon-1}}{1-\gamma_{L,\upsilon}}$ (using $\gamma_{L,\upsilon} = e^{-\frac{1}{\beta_{L,\upsilon}}}$), the dynamic normalizing value $C_{L,n}$, and the same model-exciting signal $g_k(\nu)$ as before (3.19). The formulae for calculating $\gamma_{L,n}(\nu)$ and $C_{L,n}$ are derived in the Appendix. The transfer function of the long-term digital filtering process as described by (3.22) is depicted in Fig. 3.2 as $H_L(f, n)$. Analog to (3.21) a recursive function with the same behavior as (3.22) can be defined as

$$c_{L,k}(n) = (1 - \gamma_{L,n-1}) \cdot g_k(n-1) + \gamma_{L,n-1} \cdot c_{L,k}(n-1) \tag{3.23}$$

with the same initial value $c_{L,k}(0) = \frac{1}{K}$, the proof can be found in the Appendix. Figure 3.5 shows the block diagram of the infinite impulse response (IIR) digital filter, which captures the long-term memory and illustrates the updating process inherent in (3.23). The long-term transfer function $H_L(f, n)$ is plotted in Fig. 3.3 as (- - -) for $n = 1$ and as (——) for $n = N = 192$, respectively. Inspection of Fig. 3.3 reveals an initially moderate low-pass characteristic, which becomes much sharper when the

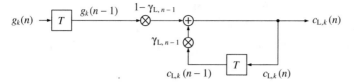

**Fig. 3.5** Block diagram of the time-variant first-order infinite impulse response length (IIR) filter with transfer function $H_L(f, n)$, which captures the long-term memory equivalent to (3.22) or (3.23). Elements $\boxed{T}$ denote a delay of one trial. At the adder element $\oplus$, updating of the weighted output $\gamma_{L,n-1} \cdot c_{L,k}(n-1)$ of the last trial $n-1$ with the weighted input $(1 - \gamma_{L,n-1}) \cdot g_k(n-1)$ of the last trial results in the current output $c_{L,k}(n)$. Note that via $\gamma_{L,n-1} \cdot c_{L,k}(n-1)$, all preceding inputs (and outputs) influence the current output

number of trials increases. The behavior of the dynamic long-term time value $\beta_{L,n}$, which controls the time-variant characteristics of $H_L(f, n)$, is following

$$\beta_{L,n} = \begin{cases} e^{-(\frac{1}{\tau_1} \cdot 1 + \frac{1}{\tau_2})}, & \text{if } n \leq 0 \\ e^{-(\frac{1}{\tau_1} \cdot n + \frac{1}{\tau_2})}, & \text{otherwise,} \end{cases} \tag{3.24}$$

with time constants $\tau_1$ and $\tau_2$ controlling the speed of transition from reliance on prior assumptions to experience. The time value $0 \leq \beta_{L,n} < \infty$ satisfies $\beta_{L,n} > \beta_S$. The effect of this dynamic formulation of $\beta_{L,n}$ is further illustrated in Fig. 3.6, which shows the values of $\beta_{L,n}$ and the corresponding forgetting factor $\gamma_{L,n} = e^{-\frac{1}{\beta_{L,n}}}$ for trials $n \in \{1, \ldots, N\}$ with $N = 192$.

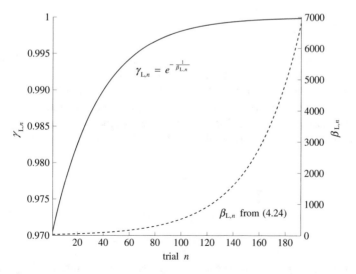

**Fig. 3.6** The dynamics of the coefficients $\gamma_{L,n}$ (———) and $\beta_{L,n}$ (- - -) over trials $n = 1, \ldots, N$, with $N = 192$

### 3.4.3  Alternation Expectation

Finally, the DIF model comprises a count function capturing alternations

$$c_{A,k}(n) = \frac{1}{C_A} \sum_{\nu=n-4}^{n-1} \gamma_{A,n-\nu} \cdot g_k(\nu),$$  (3.25)

with some normalizing constant $C_A$ and the same model-exciting signal $g_k(\nu)$ as before (3.19). In contrast to the short- and long-term memory IIR filters, which both have a low-pass characteristic, this finite impulse response (FIR) filter reveals a high-pass characteristic. Its transfer function is plotted in Fig. 3.3 as (·····). Alternation expectation is assumed to depend on the visual working memory (vWM (Baddeley 2003)), which can maintain representations of only around four objects at any given moment (Cowan 2005). Figure 3.7 shows the block diagram of this fourth-order finite impulse response filter. The coefficients $\gamma_{A,n-\nu}$ and the probability normalizing constants $C$ and $C_A$ are now described in detail. In order to reduce the complexity of the model by keeping the number of independent parameters to a minimum, only the coefficient $\gamma_{A,2}$ is chosen freely within the range $0 \le \gamma_{A,2} \le \gamma_{A,max}$, with $\gamma_{A,max} = 1$ as in Kolossa et al. (2013). The effect of the multiplicative constant $C_A$ and additive constant $C$ (as shown in Fig. 3.2) is merely to ensure $0 \le c_{A,k}(n) + \frac{1}{C} \le 1$. The remaining filter coefficients depend on $\gamma_{A,2}$ and are set following $\gamma_{A,1} = -\gamma_{A,2}$, $\gamma_{A,4} = \gamma_{A,max} - \gamma_{A,2}$ and $\gamma_{A,3} = -\gamma_{A,4}$. The normalizing constants have to be set according to $C_A = \gamma_{A,max} + \gamma_{A,2} + \gamma_{A,4} = 2\gamma_{A,max}$ and $C = \frac{\gamma_{A,max} + \gamma_{A,2} + \gamma_{A,4}}{\gamma_{A,max}} = 2$.

### 3.4.4  Explanatory Notes

There are two important differences to Squires et al. (1976): (1) The DIF model comprises *two* terms with different time parameters $\beta_S$ and $\beta_{L,n}$, with one accounting for the short-term memory (as in (3.2)), the other one accounting for a dynamically adapted long-term memory. (2) Both short- and long-term memory recall all past events, meaning $N_{depth} \to \infty$. Note that the effective memory length is determined by the time parameters. Moreover, the role of negative trial indices in the DIF model

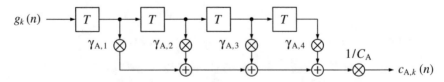

**Fig. 3.7** Block diagram of the fourth-order finite impulse response (FIR) filter $H_A(f)$. Elements $\boxed{T}$ represent a delay of one trial, the multipliers $\gamma_{A,1}$, $\gamma_{A,2}$, $\gamma_{A,3}$, and $\gamma_{A,4}$ compose the filter coefficients, and $C_A$ constitutes a normalizing constant

is to define the initial observation probability distribution, which is $P_k(1) = \frac{1}{K}, \forall k \in \{1, \ldots, K\}$ in (3.18), reflecting that the subjects were not informed about the actual relative frequencies of events over a block of trials. Equivalently, this is achieved by setting initial conditions of the delay elements of the filters.

In summary, the DIF model expresses both a long-term memory contribution and a short-term memory contribution by exponential-decay processes, with uniform initial prior observation probabilities, and a contribution of alternation expectation. Though there are similarities between (3.1) and (3.18), the model of Squires et al. (1976) uses information about the experimental design ($P_k$), which was unavailable to the subjects. In contrast, the DIF model uses only the information contained in the stimulus sequence as observed by subjects, and it starts with a uniform initial observation probability distribution $P_k(1) = \frac{1}{K}, \forall k \in \{1, \ldots, K\}$, regardless of the actual relative frequencies of events over a block of trials. Finally, it should be noted that the DIF model yields a conceptually well-defined observation probability $P_k(n) \in [0, 1]$, as opposed to a not-normalized expectancy as in Squires et al. (1976).

### 3.4.5  Surprise Based on the DIF Model

Analogous to the models taken from literature (see Sect. 3.3.4), the surprise framework (see Sect. 1.4) is applied to the DIF model as well. Thus, predictive surprise

$$I_P(n) = -\log_2 P_{k=o(n)}(n) \tag{3.26}$$

is calculated based on the observation probability $P_{k=o(n)}(n)$ (3.18) of $o(n) = k$ on trial $n$, while the change from the observation probability distributions $P_\mathcal{K}(n)$ (before observation $o(n)$) to $P_\mathcal{K}(n+1)$ (after $o(n)$ has been observed) is measured with postdictive surprise (1.18)

$$I_B(n) = D_{KL}\left(P_\mathcal{K}(n) \,\|\, P_\mathcal{K}(n+1)\right) = \sum_{k \in \mathcal{K}} P_k(n) \log\left(\frac{P_k(n)}{P_k(n+1)}\right). \tag{3.27}$$

### 3.4.6  DIF Model Parameter Training

The values for the free model parameters, namely $\alpha_L$ (3.18), $\tau_1$ and $\tau_2$ (both (3.24)), $\beta_S$ (for (3.20)), $\alpha_A$ (3.18) and $\gamma_{A,2}$ (3.25) are trained on the data from the first three blocks $b \in \{1, 2, 3\}$ of each experimental condition, which avoids circularity (see Sect. 1.3). As the DIF model enters the model space with postdictive and predictive surprise as response functions, the parameters are optimized for each kind of surprise

**Table 3.2** The ranges of the free model parameters and the starting values as found in (Kolossa et al. 2013)

| Parameter | Min | Max | Initial value |
| --- | --- | --- | --- |
| $\alpha_L$ | 0.5 | 0.9 | 0.83 |
| $\tau_1$ | 10 | 100 | 33.64 |
| $\tau_2$ | 0.1 | 1 | 0.27 |
| $\beta_S$ | 1 | 10 | 1.82 |
| $\alpha_A$ | 0.001 | 0.1 | 0.05 |
| $\gamma_{A,2}$ | 0.5 | 1 | 0.94 |

Note that $\alpha_S$ is not a free parameter due to the restriction $\alpha_S = 1 - \alpha_L - \alpha_A$, and thus not optimized independently

separately. Testing the whole range of possible combinations of the parameters with a reasonable resolution is computationally too expensive (testing 100 values per parameter would result in a runtime of several years). Therefore, the parameters are optimized by means of a pattern search algorithm (Ingber 1996; Kolda et al. 2006) with the absolute value of the variational free energy on the group level $|F_m|$ as objective function (see Sect. 2.4.4 for details on group studies). The specification of the data vectors and design matrices is exactly as specified in Sect. 3.5, but with the first three blocks ($b \in \{1, 2, 3\}$) instead of the last three blocks ($b \in \{4, 5, 6\}$). Note that the *same sample* of model parameters is used for all subjects. Table 3.2 gives an overview over the parameter ranges and starting values, which were set to those found in (Kolossa et al. 2013).

## 3.5   Specification of the Design Matrices for Model Estimation and Selection in the Oddball Task

As introduced in Chap. 2, the parametric empirical Bayes framework (see Sect. 2.4) is used for model estimation (see Sect. 2.4.2), and model selection rests upon posterior model probabilities for group studies $P(m|\mathbf{y})$ derived from the variational free energy $F_m$ (see Sect. 2.4.4). The model space $\mathcal{M}$ consists of all combinations of observer models and surprise along with the null model (NUL), namely $m \in \mathcal{M} = \{\text{SQU}, I_B(\text{MAR}), I_P(\text{MAR}), I_B(\text{OST}), I_B(\text{DIF}), I_P(\text{DIF}), \text{NUL}\}$. The sequence $i_m(n)$ captures the surprise from model $m$. This section describes the composition of the model-independent data vector $\mathbf{y}_{\ell,c}$ and the model-specific design matrices $\mathbf{X}^{(1)}_{m,\ell,c}$ for each subject $\ell = \{1, \ldots, 16\}$ and experimental condition $c \in \mathcal{C} = \{[0.5, 0.5], [0.3, 0.7]\}$ (see Sect. 3.2 for details on the experimental setup). As stated in Sect. 1.3, the data sets for the parameter training of the DIF model have to be separate from the data sets for model selection. Thus, only the blocks $b \in \{4, 5, 6\}$ of each experimental condition $c$ are used for model estimation and selection. The two-level GLM (see Sect. 2.3.1) is used with a single predictor per trial, simplifying (2.16) to

$$\mathbf{y}_{\ell,c} = \mathbf{x}_{m,\ell,c}^{(1)} \theta_{m,\ell,c}^{(1)} + \boldsymbol{\epsilon}_{m,\ell,c}^{(1)}$$
$$\theta_{m,\ell,c}^{(1)} = \epsilon_{m,\ell,c}^{(2)}. \tag{3.28}$$

The vector $\mathbf{y}_{\ell,c} = [\mathbf{y}_{\ell,c,b=4}^T, \mathbf{y}_{\ell,c,b=5}^T, \mathbf{y}_{\ell,c,b=6}^T]^T$ with $\mathbf{y}_{\ell,c,b} = [y_{\ell,c,b}''(n = 1), \ldots, y_{\ell,c,b}''(n = N_{\ell,c,b})]^T$ contains the P300 amplitudes which were normalized to zero mean and unit variance following

$$y_{\ell,c,b}''(n) = \frac{y_{\ell,c,b}(n) - \overline{y}_{\ell,c,b}}{\sigma_{y_{\ell,c,b}}}, \tag{3.29}$$

with mean

$$\overline{y}_{\ell,c,b} = \frac{1}{N_{\ell,c,b}} \sum_{n=1}^{N_{\ell,c,b}} y_{\ell,c,b}(n) \tag{3.30}$$

and standard deviation

$$\sigma_{y_{\ell,c,b}} = \sqrt{\frac{1}{N_{\ell,c,b} - 1} \sum_{n=1}^{N_{\ell,c,b}} (y_{\ell,c,b}(n) - \overline{y}_{\ell,c,b})^2}. \tag{3.31}$$

Note that while each subject perceived $n \in \{1, \ldots, 192\}$ stimuli on each block $b$, trials during which an artifact occurred or the wrong behavioral response was elicited are excluded from further analyses, leaving $n \in \{1, \ldots, N_{\ell,c,b}\}$ trials. Analogous to (3.29), (3.30), and (3.31), the surprise values $i_{m,\ell,c,b}(n)$ are normalized to $i_{m,\ell,c,b}''(n)$ with zero mean and unit variance before constituting $\mathbf{x}_{m,\ell,c}^{(1)} = [\mathbf{i}_{m,\ell,c,b=4}^T, \mathbf{i}_{m,\ell,c,b=5}^T, \mathbf{i}_{m,\ell,c,b=6}^T]^T$ with $\mathbf{i}_{m,\ell,c,b} = [i_{m,\ell,c,b}''(n = 1), \ldots, i_{m,\ell,c,b}''(n = N_{\ell,c,b})]^T$. After having $\mathbf{y}_{\ell,c}$ and $\mathbf{x}_{m,\ell,c}^{(1)}$ fully specified, model estimation commences as described in Sect. 2.4.2. Note that for the NUL model $\mathbf{i}_{\mathrm{NUL},\ell,c,b} = [1, \ldots, 1]^T \in \mathbb{R}^{N_{\ell,c,b}}$ is an all-one vector as in Sect. 2.5.3.4.

The obtained experimental condition-specific variational free energies $F_{m,\ell,c}$ are summed up to yield $F_{m,\ell} = \sum_{c \in C} F_{m,\ell,c}$. Based on fixed effects assumptions (see Sect. 2.4.4), the group-level variational free energy is then given as $F_m = \sum_{\ell=1}^{L} F_{m,\ell}$. Finally, posterior model probabilities $P(m|\mathbf{y}) = \frac{e^{F_m}}{\sum_{m \in \mathcal{M}} e^{F_m}}$ are calculated. The conditional means of the first-level parameters $\theta_{m,\ell,c}^{(1)}$ of the posterior densities $\mu_{\theta_{m,\ell,c}|\mathbf{y}}^{(1)}$ (2.51) are used as maximum a posteriori point estimates of the parameters for the model fitting for Fig. 3.11.

## 3.6   Results

This section first presents the conventional ERP analyses in the form of grand-average ERP waveforms for frequent and rare events, as well as sequence-specific ERP waveforms and the signal-to-noise ratio estimates. Next, the model-based trial-by-trial analyses present the posterior model probabilities for the whole model space. A detailed discussion of the behavior of the DIF model in comparison to the SQU and MAR models taken from literature concludes this section.

### 3.6.1   Conventional ERP Analyses

#### 3.6.1.1   Grand-Average and Sequential ERP Waveforms

Figure 3.8a shows grand-average ERP waveforms that were obtained for the frequent event (——) and for the rare event (- - -) at Pz in the [0.3, 0.7] experimental condition. The grand-average ERP waveforms show a clear dependency on event probability in the P300 typical time interval around 300 ms. Figure 3.8b depicts third-order sequence effects on ERP waveforms at Pz that were obtained in the [0.5, 0.5] experimental condition. Note that ERP waveforms in response to sequences of four successive stimuli are illustrated. The sequences are labeled in temporal order (trial $n - 3$, trial $n - 2$, trial $n - 1$, trial $n$), with $a$ signifying one particular event and $b$ the other one. The area of P300 peak amplitudes is magnified for a better discrimination between the different curves. For example, $aaaa$ indicates event $a$ being repeated across four consecutive trials (shown as (- - -)), whereas $bbba$ represents the presentation of event $a$ after having event $b$ repeated across the three immediately preceding trials (shown as (- - -)). Note further that the two solid traces, originating from the $abaa$ (——) and the $baba$ (——) sequences, show reversed P300 amplitudes. Specifically, for the single -$b$- sequence $abaa$ (——), the P300 waveform lies within those from dual -$bb$- sequences, whereas for the dual -$bb$- sequence $baba$ (- - -), the P300 waveform appears indistinguishable from the waveforms from single -$b$- sequences. As further detailed in the discussion later on, this amplitude reversal is attributed to the disconfirmation of alternation expectation in the $abaa$ (——) sequence, as well as to the confirmation of alternation expectation in the $baba$ (——) sequence. Note that the sequence effect in the [0.3, 0.7] experimental condition is superimposed with the probability effect and that, naturally, there is no probability effect in the [0.5, 0.5] experimental condition. These factors will be shown in depth later on in Sect. 3.6.2.

**(a)**

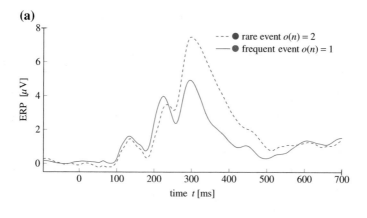

**(b)**

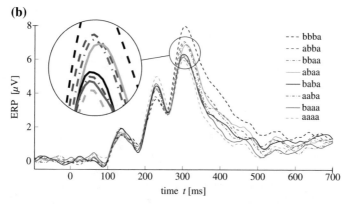

**Fig. 3.8** Grand-average and sequence-specific ERP waveforms. **a** Grand-average ERP waveforms at Pz show the probability effect on P300 amplitudes in the [0.3, 0.7] experimental condition. The grand-average ERP waveform of the rare event (- - -) lies above that of the frequent event (——) in the P300 typical time interval around 300 ms. **b** Sequence-specific ERP waveforms at Pz show the sequence effect on P300 amplitudes in the [0.5, 0.5] experimental condition for sequences of four successive stimuli; $a$ signifies one particular event ($b$ the other one). The area of P300 peak amplitudes is magnified for a better discrimination between the different curves

### 3.6.1.2 Signal-to-Noise Ratio Estimates

Table 3.3 shows subject-specific signal-to-noise ratio estimates $\widehat{\mathrm{SNR}}\,[\mathrm{dB}]$ (1.17) for the frequent (●) and rare (●) event types in both experimental conditions $c \in \mathcal{C}$. The experimental condition-specific average $\widehat{\mathrm{SNR}}\,[\mathrm{dB}]$ values over both types of events are shown in the right column of the respective experimental condition. The average $\widehat{\mathrm{SNR}}\,[\mathrm{dB}]$ values over both experimental conditions and event types are presented in the rightmost column. The bottom row shows the average values for the respective columns highlighted in bold face. For the [0.5, 0.5] experimental condition, the average $\widehat{\mathrm{SNR}}\,[\mathrm{dB}]$ values are nearly identical with $\widehat{\mathrm{SNR}}_{k=1}\,[\mathrm{dB}] = 1.70\,\mathrm{dB}$ and $\widehat{\mathrm{SNR}}_{k=2}\,[\mathrm{dB}] = 1.83\,\mathrm{dB}$, respectively, which coincides with equally proba-

**Table 3.3** Subject-specific and average signal-to-noise ratio estimates $\widehat{\text{SNR}}$ [dB] (1.17) for the P300 at electrode Pz

| Condition | Signal-to-noise ratio estimate [dB] | | | | | | Both |
|---|---|---|---|---|---|---|---|
| | [0.5, 0.5] | | | [0.3, 0.7] | | | |
| Subject # | ● | ● | ● | ● | ● | ● | ● |
| 1 | 1.90 | 3.02 | 2.46 | 0.27 | 2.18 | 0.90 | 1.68 |
| 2 | 0.38 | 0.90 | 0.64 | 0.13 | 0.74 | 0.33 | 0.49 |
| 3 | 1.51 | 2.90 | 2.21 | 1.70 | 3.08 | 2.16 | 2.19 |
| 4 | 6.38 | 3.67 | 5.03 | 2.10 | 7.23 | 3.78 | 4.38 |
| 5 | 3.22 | 3.47 | 3.35 | 3.25 | 4.88 | 3.79 | 3.57 |
| 6 | 3.30 | 2.76 | 3.03 | 2.73 | 3.93 | 3.13 | 3.08 |
| 7 | 2.97 | 4.10 | 3.53 | 2.31 | 4.25 | 2.96 | 3.24 |
| 8 | 0.00 | 0.02 | 0.01 | 0.03 | 0.16 | 0.08 | 0.04 |
| 9 | 0.24 | 1.34 | 0.79 | 0.54 | 1.34 | 0.81 | 0.80 |
| 10 | 0.82 | 0.36 | 0.59 | 0.02 | 1.07 | 0.37 | 0.48 |
| 11 | 0.75 | 1.54 | 1.15 | 0.25 | 2.88 | 1.13 | 1.14 |
| 12 | 0.46 | 0.05 | 0.26 | 0.00 | 1.23 | 0.41 | 0.33 |
| 13 | 0.00 | 0.19 | 0.09 | 0.16 | 0.03 | 0.12 | 0.10 |
| 14 | 0.43 | 0.46 | 0.44 | 0.42 | 0.38 | 0.41 | 0.42 |
| 15 | 2.14 | 3.26 | 2.70 | 1.00 | 4.21 | 2.06 | 2.38 |
| 16 | 2.70 | 1.27 | 1.99 | 0.80 | 3.00 | 1.53 | 1.76 |
| All | **1.70** | **1.83** | **1.77** | **0.98** | **2.54** | **1.50** | **1.63** |

The $\widehat{\text{SNR}}$ [dB] values are shown separately for the frequent (●) and rare (●) event types in both experimental conditions $c \in C$ and as averages over event types (●). The experimental condition-specific average $\widehat{\text{SNR}}$ [dB] values over both types of events are shown in the right column of the respective experimental condition. The average $\widehat{\text{SNR}}$ [dB] values over both experimental conditions and event types are presented in the rightmost column. The bottom row shows the average estimated signal-to-noise ratios for the respective columns

ble events. In the [0.3, 0.7] experimental condition $\widehat{\text{SNR}}_{k=1}$ [dB] $= 0.98$ dB and $\widehat{\text{SNR}}_{k=2}$ [dB] $= 2.54$ dB, which conforms to larger ERP amplitudes for the rare event as shown in Fig. 3.8a. The minimum $\widehat{\text{SNR}}$ [dB] values are 0 dB, and the overall average is $\widehat{\text{SNR}}$ [dB] $= 1.63$ dB. Taking the simulation results for group-level model selection from Sect. 2.6 into account, these signal-to-noise ratios indicate that the 576 trials per subject, which remain after splitting off the 576 trials for the DIF model parameter training, allow for reliable model selection results. Specifically Fig. 2.6 shows that with a signal-to-noise ratio larger than 0 dB more than 250 trials per subject are sufficient for reliable model selection results.

### 3.6.2 Model-Based Trial-by-Trial Analyses

#### 3.6.2.1 Posterior Model Probabilities and Trained Parameters of the DIF Model

Table 3.4 shows the posterior model probabilities for the whole model space $m \in \mathcal{M} = \{$SQU, $I_B(\text{MAR})$, $I_P(\text{MAR})$, $I_B(\text{OST})$, $I_B(\text{DIF})$, $I_P(\text{DIF})$, NUL$\}$. Predictive surprise based on the DIF model is clearly favored with *very strong* evidence $P(I_P(\text{DIF})|\mathbf{y}) > 0.99$ (Kass and Raftery 1995; Penny et al. 2004). There is not even weak evidence for any other model with all posterior model probabilities $P(m|\mathbf{y}) < 0.01$. Note that if $I_P(\text{DIF})$ is excluded from the model space, the SQU model is superior with $P(\text{SQU}|\mathbf{y}) > 0.99$, making it the unrivaled second best model (see bottom row of Table 3.4).

Table 3.5 shows the parameters of the DIF model which were trained on the first half of the data and used for calculating the posterior model probabilities in Table 3.4. The high value of $\alpha_L = 0.82$ shows that the observation probability mainly follows the long-term memory. The identified values for $\tau_1$ and $\tau_2$ yield $\beta_{L,1} = 33$ and $\beta_{L,192} = 6803$, which was further illustrated in Fig. 3.6. With a short-term memory time constant of $\beta_S = 1.49$ and a weight of $\alpha_S = 0.13$ the influence of recent events to the observation probability is captured. Whereas the weight of the filter modeling alternation expectation $\alpha_S = 0.05$ appears to be small, the importance of this contribution will be shown shortly. Postdictive surprise based on the DIF model does not receive any support from the data, and the optimized parameters $\alpha_L = 0.5$ and $\tau_1 = 100$ are boundary values from the allowed parameter ranges, suggesting results based on chance. Thus, the behavior of the DIF model with these parameters is not further discussed. Postdictive surprise based on the MAR model does not receive any support as well, which confirms the results presented by Mars et al. (2008). In the following, the behavior of the SQU, MAR, and DIF models are described in more detail. Solely predictive surprise is used as response function, as it proved clearly superior for the DIF model (see Table 3.4) and was originally proposed for the MAR model in (Mars et al. 2008).

**Table 3.4** Posterior model probabilities for the whole model space and the model space with $I_P(\text{DIF})$ removed

|            | SQU    | $I_B(\text{MAR})$ | $I_P(\text{MAR})$ | $I_B(\text{OST})$ | $I_B(\text{DIF})$ | $I_P(\text{DIF})$ | NUL    |
|------------|--------|--------|--------|--------|--------|--------|--------|
| P(m\|y)   | <0.01  | <0.01  | <0.01  | <0.01  | <0.01  | **>0.99** | <0.01  |
| P(m\|y)   | **>0.99** | <0.01  | <0.01  | <0.01  | <0.01  |        | <0.01  |

**Table 3.5** The optimized model parameters

|                  | $\alpha_L$ | $\tau_1$ | $\tau_2$ | $\alpha_S$ | $\beta_S$ | $\alpha_A$ | $\gamma_{A,2}$ |
|------------------|--------|------|------|------|------|------|------|
| $I_P(\text{DIF})$ | 0.82   | 35.9 | 0.29 | 0.13 | 1.49 | 0.05 | 0.90 |
| $I_B(\text{DIF})$ | 0.50   | 100  | 0.57 | 0.48 | 2.47 | 0.02 | 0.93 |

Note that $\tau_1 = 35.9$ and $\tau_2 = 0.29$ yield $\beta_{L,1} = 33$ and $\beta_{L,192} = 6803$, respectively

#### 3.6.2.2   Trial-by-Trial Model Behavior

Fig. 3.9 illustrates the trial-by-trial behavior of the SQU ((3.1), (- - -)), MAR ((3.7), (---)), and DIF ((3.18), (—)) models for trials $n \in \{1, \ldots, 100\}$ for a random stimulus sequence in the $[0.3, 0.7]$ experimental condition. The stimulus sequence is indicated above and below the curves in all panels by (●) signifying a frequent event and by (●) signifying a rare event.

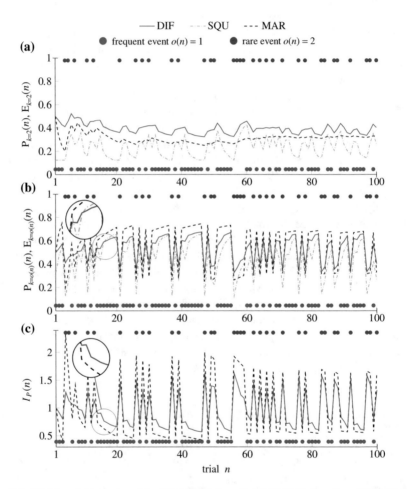

**Fig. 3.9** Trial-by-trial behavior of the SQU ((3.1), (- - -)), MAR ((3.7), (- - -)), and DIF ((3.18), (—)) models in the $[0.3, 0.7]$ experimental condition for trials $n \in \{1, \ldots, 100\}$. The stimulus sequence is indicated above and below the curves by (●) signifying a frequent event and by (●) signifying a rare event. **a** Observation probabilities $P_{k=2}(n)$ and expectancies $E_{k=2}(n)$ for the rare stimulus (●). **b** Observation probabilities $P_{k=o(n)}(n)$ and expectancies $E_{k=o(n)}(n)$ for the actually presented stimulus $k = o(n)$. **c** Predictive surprise values of the MAR (3.15) and DIF (3.26) models

The upper panel shows the observation probability $P_{k=2}(n)$ and expectancy $E_{k=2}(n)$ of seeing stimulus $k = 2$ (●) on trial $n$. While the MAR model shows dynamic behavior for the first trials, it becomes less dynamic starting around $n = 20$ trials and is nearly constant from around $n = 60$ trials on. Note that it approximates the relative stimulus frequency $P_{k=2} = 0.3$ quite well. The SQU model is very dynamic, which is mostly driven by the short-term memory and alternation expectation. For the DIF model, $P_{k=2}(n)$ shows the effect of the long-term memory contribution by a slow decrease from the initial prior to a lower probability, but it stays on a level of higher uncertainty than the MAR model. The contributions of the short-term memory and alternation expectation are clearly visible by an oscillation around the long-term memory contribution.

The middle panel shows plots of the expectancy $E_{k=o(n)}(n)$ and the observation probability $P_{k=o(n)}(n)$ of observing *the actually occurring stimulus* $k = o(n)$ on trial $n$. The transition from the upper panel to the middle panel illustrates that the observation probability is traced as a distribution $P_\mathcal{K}(n)$ for all possible events $k \in \mathcal{K}$ simultaneously over all trials, and that at the moment of observing a new stimulus $k = o(n)$ only the corresponding observation probability of that event $k$ is taken from the distribution. It is obvious that the DIF model is smoother over trials than the SQU model, but not as static as the MAR model, which becomes almost binary over increasing trial number $n$. Furthermore, the importance of alternation expectation becomes most apparent whenever successive stimuli are identical. One such area is magnified in order to make this effect more apparent. It can be seen that the probability estimated by the MAR model increases monotonously, while the probabilities estimated by the SQU and DIF models first decrease when the alternation expectation is violated. This area additionally emphasizes how the DIF model combines properties of the SQU and MAR models, because Mars et al's model does not account at all for the well-documented sequence effects (Squires et al. 1976). Note that expectancy $E_k(n)$ is *not* a normalized probability, so contrary values from the upper and middle panel do not sum to one ($E_{k=1}(n) + E_{k=2}(n) \neq 1$).

The lower panel shows predictive surprise (1.20) based on the DIF and MAR models. For both models, the main property of predictive surprise becomes apparent: If an event with a high probability is observed, it causes small surprise, while the observation of an event with a low probability is accompanied by increasing surprise values. The tendency of the DIF model toward less extreme values than the MAR model, which was shown in the upper and middle panels, clearly persists for predictive surprise as well. The effect of alternation expectation further discriminates the two models as shown in the middle panel. One such distinctive area is magnified for better perceptibility. Predictive surprise based on the MAR model decreases monotonously, while predictive surprise based on the DIF model increases slightly before it decreases as well.

### 3.6.2.3    Sequence-Specific ERPs and Fitting Parameters

The model-based predictive surprise and expectancy values are fitted following (2.10) to yield the clean data estimate $\widehat{s}(n)$ for further visualization. As described in Sect. 3.5, the surprise values and the P300 amplitudes were normalized to zero mean and unit variance for model estimation (3.29). The estimated means of the first-level parameter densities of the general linear model (3.28) were used as point estimates for the fitting parameters $\theta_{\ell,c}^{(1)}$ as stated in Sect. 3.5 (see page 27 in Sect. 2.4.2 for a detailed description). Figure 3.10 shows the subject-specific fitting parameters for the DIF model (▮▮), the SQU model (▮▮), and the MAR model (▮▮) as bar plots for

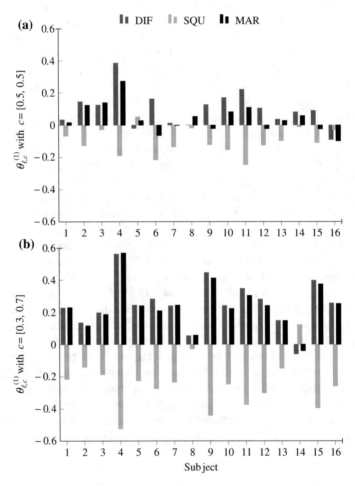

**Fig. 3.10** Point estimates of the subject- and experimental condition-specific first-level parameters $\theta_{\ell,c}^{(1)}$ for the DIF (▮▮), SQU (▮▮), and MAR (▮▮) models. **a** Parameters for the $c = [0.5, 0.5]$ experimental condition. **b** Parameters for $c = [0.3, 0.7]$

**Table 3.6** Fitting parameter point estimates averaged over subjects ($\theta_c^{(1)}$) and experimental conditions ($\theta^{(1)}$) for the DIF, SQU, and MAR models

| | $\theta_c^{(1)}$ with $c = [0.5, 0.5]$ | $\theta_c^{(1)}$ with $c = [0.3, 0.7]$ | $\theta^{(1)}$ |
|---|---|---|---|
| DIF | 0.0989 | 0.2504 | 0.1747 |
| SQU | −0.1022 | −0.2425 | −0.1724 |
| MAR | 0.0422 | 0.2358 | 0.1390 |

the [0.5, 0.5] experimental condition (upper panel) and the [0.3, 0.7] experimental condition (lower panel). Inspection of Fig. 3.10 reveals a large variation over subjects.

Table 3.6 shows the experimental condition-specific parameters after averaging over subjects ($\theta_c^{(1)}$) and after additional averaging over the experimental conditions ($\theta^{(1)}$). Apart from the algebraic sign, the fitting parameters of the DIF and SQU models are very similar with a slightly bigger absolute value of $\theta^{(1)}$ for the DIF model. The values for the MAR model are prominently smaller than those of the DIF and SQU models, especially in the [0.5, 0.5] experimental condition, which is most likely credited to its static behavior and insensitivity to the stimulus sequence. Overall, these results confirm the superiority of the DIF model, and the SQU model as second best model. Notice that the comparison of the fitting parameters is sound, because the data and regressors were normalized prior to model estimation (Hoijtink 2012, p. 17), and that the majoritarian negative algebraic signs of the parameters for the SQU model are *not* penalized when calculating the variational free energy for model selection. Anyway, the negative algebraic signs call for some attention, as they give clear data-based evidence to the merit of predictive surprise as response function.

Figure 3.11 shows the tree diagrams of the ERPs ((—●—), (—●—), (—●—)), the fitted predictive surprise values for the DIF model (—●—) as well as the MAR model (··●··), and the fitted expectancy values of the SQU model (-●-) as a function of the preceding stimulus sequences for the different experimental conditions. To derive these trees, the normalized trial-by-trial P300 amplitudes and the model-based estimates were averaged according to eight third-order stimulus sequences (denoted as *aaaa*, *baaa*, *abaa*, *aaba*, *bbaa*, *abba*, *baba*, *bbba*), four second-order stimulus sequences (*aaa*, *baa*, *aba*, *bba*), and two first-order sequences (*aa*, *ba*). Please note that in the [0.3, 0.7] experimental condition the symbol *a* denotes the rare event (●) yielding the corresponding curve (—●—), while *b* signifies the frequent event (●) with the associated curve (—●—). In the [0.5, 0.5] experimental condition, sequential analysis could be collapsed across the two types of events since both stimuli were equally probable and task-relevant, meaning that *a* represents either event (●) *or* (●), while *b* denotes the other one, producing the curve (—●—). The DIF and SQU models are both capable of estimating the outer branches and inner tree structure quite well, but the SQU model fans out too much for higher order effects for the frequent stimulus (●) in the [0.3, 0.7] experimental condition. For the MAR model, higher order effects are nonexistent in the [0.5, 0.5] experimental condition and only hinted at in the [0.3, 0.7] experimental condition.

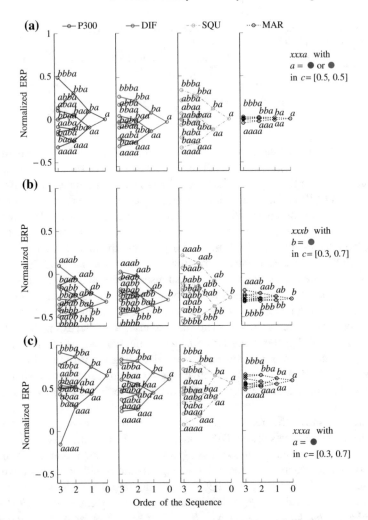

**Fig. 3.11** Tree diagrams of normalized measured ERPs and model-based fitted ERP amplitudes as a function of the sequence of preceding stimuli. The stimulus sequences are labeled and related sequences are connected by *lines*. **a** For both stimuli $a \in \{\bullet, \bullet\}$ on experimental condition [0.5, 0.5]. **b** For the frequently occurring stimulus $b = \bullet$ on experimental condition [0.3, 0.7]. **c** For the rarely occurring stimulus $a = \bullet$ on experimental condition [0.3, 0.7]

## 3.7   Summary and Discussion

This chapter introduced state-of-the-art models and a newly derived observer model which keep track of the probability distributions over frequent and rare events in an oddball task. The SQU (Squires et al. 1976), MAR (Mars et al. 2008), and OST (Ostwald et al. 2012) models were taken from literature, and the model proposed by Squires et al. (1976) has been presented in a purely computational form. Properties of the SQU and MAR models were fused into the new digital filtering (DIF) (3.18) model, which was introduced here. A formal framework of Bayesian updating and predictive surprise was used to relate the probability distributions based on the observer models to P300 amplitude fluctuations, and Bayesian model selection was used to choose the best model. The P300 amplitudes were obtained at electrode Pz in an oddball task. Conventional ERP analyses revealed an effect of the event probability and the immediately preceding stimulus sequence, which is consistent with earlier reports. The model-based analyses showed superiority of predictive surprise based on the DIF model with very strong statistical evidence. Postdictive surprise did not receive any support from the data.

**Detailed Discussion of the DIF Model**

The DIF model possesses important advantages over previous models of P300 amplitude fluctuations. It relies completely on mathematical notations and definitions, unlike the notions of expectancy (Squires et al. 1976), global versus local probability (Squires et al. 1976), temporal probability (Gonsalvez and Polich 2002) or context updating (Donchin and Coles 1988). It is a formal model, akin to the MAR model (Mars et al. 2008), but it offers a more extensive explanation of trial-by-trial P300 amplitude fluctuations. The advantage of the DIF model over the MAR model stems, in large parts, from the non-negligible contribution of the short-term memory traces to the observation probability, as evidenced by the low sensitivity of the MAR model to the sequential effects on P300 amplitudes (see Figs. 3.9 and 3.11).

The SQU model (Squires et al. 1976) can be considered as a precursor of the DIF model insofar as it comprises memory for event frequencies within the prior stimulus sequence (equivalent to the short-term trace), event probabilities (loosely related to the long-term trace), and alternation expectancy. An important distinction is, however, that the long-term contribution to the DIF model is *learned* by counting observed events in continuously larger samples and not simply *made known* to the model. Thus, observation probabilities are constantly revised while evidence is accumulating, and the DIF model is a pure model of learned statistical parameters, rather than a mixture of learned and objective parameters such as the SQU model.

The optimal short-term time constant $\beta_S$ approximates the value of 1.5, while $\beta_{L,n}$ varies as a function of $n$ (3.24), with $\gamma_{L,n}$ gradually approximating the value of one. Note that $\beta_{L,n}$ (and hence $\gamma_{L,n}$) is relatively low during early trials when compared to late trials (Fig. 3.6), which means that the long-term character of the long-term memory contribution to the observation probability $P_k(n)$ is gradually increasing as a function of $n$. This behavior is mirrored by the dynamics of the long-term low-pass filter as seen in Fig. 3.3. In other words, the decay half-life of the long-term trace gradually increases when the observer experiences more and more trials. The recursive formulation of the long-term memory (3.23) emphasizes this point by showing that the balance between the most recently experienced stimuli (which occurred on trial $n - 1$, weighted by $(1 - \gamma_{L,n-1})$) and the counted frequency (weighted by $\gamma_{L,n-1}$) is biased toward recent stimuli during early trials (when $\gamma_{L,n-1}$

is relatively low), but biased toward the counted frequency during late trials (when $\gamma_{L,n-1}$ is relatively high).

The alternation term in the DIF model (3.25), $c_{A,k}(n)$, is a finite impulse response (FIR) filter which resembles the alternation term in the SQU model (Squires et al. 1976). This high-pass filter searches for alternation patterns over short sequences of trials (such as those in *abab* and in *baba* sequences). The discovery of such patterns causes the DIF model to expect the completion of the pattern in the upcoming trial, an expectation which will be confirmed in the *ababa* sequence, but will be disconfirmed in the *babaa* sequence. The sensitivity of the P300 amplitudes to these patterns is shown in the sequence-specific ERP waveforms in Fig. 3.8 (see also Jentzsch and Sommer 2001; Ford et al. 2010). Alternation expectation applies even for first-order sequences, as revealed by the larger P300 amplitudes in response to *ba* sequences compared to *aa* sequences (Fig. 3.11).

## Neuropsychological Implications of the Digital Filtering Model

Given the superiority of the DIF model, the general implications of its properties are now shortly considered. To begin with it is important to view the DIF model in the context of the processing of event frequencies (Sedlmeier and Betsch 2002). In particular, the reliable encoding of the frequency with which events occur (Hintzman 1976; Underwood 1969) led to the claim that event frequency is automatically encoded in memory, placing only minimal demands on attentional resources (Hasher and Zacks 1984; Zacks and Hasher 2002). There are multiple theories of how event frequency is represented. According to multiple-trace views, a record of individual events is stored such that each attended occurrence of an event results in an independent memory trace (Hintzman 1976). In contrast, according to strength views, each attended event occurrence produces an increment in the strength of a single memory trace or a frequency counter (Alba et al. 1980). The latter corresponds to the assumptions of event frequency counters inherent in the DIF model, which are instantiated as the frequency counters ((3.20), (3.22)), which are the short-term, $c_{S,k}(n)$, and the long-term, $c_{L,k}(n)$, memory traces, respectively. These traces are of exponential-decay nature (Lu et al. 2005), differing mainly with regard to their decay half-lives. The dual decay rate assumption is compatible with the fact that short-term and long-term memory functions depend on dissociable neuronal processes (Jonides et al. 2008). Further, recent functional brain imaging data suggest different distributions of neural activities for short-term and long-term decay functions (Harrison et al. 2011). The visual working memory (vWM (Baddeley 2003)) can maintain representations of only around four objects at any given moment (Cowan 2005). This limited capacity offers a reason for the capacity-limited alternation term (3.25) in the DIF model, $c_{A,k}(n)$.

The DIF model offers a digital filtering account of multiple memory systems in the brain (Figs. 3.2 and 3.3). Specifically, the DIF model characterizes frequency memory as two digital first-order infinite impulse response (IIR) low-pass filters: One filter with an experience-invariant short-term exponential-decay function (Fig. 3.4), and another filter with an experience-dependent long-term exponential-decay function, such that the low-pass characteristics become stronger as the amount of experience increases (Fig. 3.5). This dynamic behavior suggests that the human brain relies more and more on environmental experience as it becomes available, rather than on prior assumptions, which enables it to exploit nonrandom probabilities in a chance-driven world (Barlow 2001). Moreover, vWM is conceptualized as an additional fourth-order finite impulse response (FIR) high-pass filter (Fig. 3.7). The input signal $g_k(n)$ (3.19) to all three filters is a binary representation of the stimulus sequence, with all samples prior to the first trial defining a uniform initial prior.

**Implications for Predictive Coding and the Free-Energy Principle**

The theory of variation in trial-by-trial P300 amplitudes as implied by the DIF model bears implications on the nature of cortical processing. It is in agreement with the predictive coding theory (Friston 2002; Friston 2005; Spratling 2010) and the free-energy principle (Friston 2010). Viewed from the perspective of predictive coding, predictive surprise—and hence trial-by-trial P300 amplitude—is proportional to the residual error between top-down priors and bottom-up sensory evidence. Further, the DIF model is a Bayesian model of cortical processing (Knill and Pouget 2004; Friston 2005). It represents performance in the oddball task as sequential Bayesian learning (MacKay 2003) in order to reduce predictive surprise over future observations, which conforms to the free-energy principle.

**Alternative Interpretations**

It is important to note that it cannot be determined whether the observed P300 modulations were exclusively related to predictive surprise *over sensory input*. The reason is that in the present task design the probabilities of events were mirrored on probabilities of motoric responses, as each stimulus was mapped onto a distinct motoric response. Thus, a particular stimulus also called for the respective motor program, and it is possible that the observed P300 modulations are related to predictive surprise *over motor responses*. Absolute certainty can not be attained about whether the measured P300 amplitude fluctuations are due to surprise conveyed by the visual stimuli, or whether they are related to surprise associated with the selection of a motor response, given a visual stimulus on each trial (Barceló et al. 2008; O'Connell et al. 2012).

**Next Steps**

While these analyses yield substantial evidence for the coding and computing of probability distributions in the brain, they lack neural evidence for the updating of the probability distributions from $P_K(n)$ to $P_K(n+1)$ themselves. Further, the oddball task does not provide the subjects with any kind of prior knowledge about the probability distributions over events, but the incorporation of prior knowledge is characteristic for Bayesian inference. To address both of theses issues, a new task has to be tailored to fit Bayes theorem and the scope of the analyzed data has to be increased in search for evidence for the actual updating of the probability distributions.

# Chapter 4
# Bayesian Inference and the Urn-Ball Task

This chapter introduces a Bayesian observer model and an urn-ball task which is tailored to fit Bayes' theorem and equips the subjects with prior knowledge about the distributions over the random variables contained in the task. The Bayesian observer model adjusts internal beliefs about hidden states in the environment and predictions about observable events. The scope of the analyzed data is extended to the complete late positive complex (P3a, P3b, Slow Wave) and the N250. It starts with a brief overview on the Bayesian observer model and the urn-ball task and their relation to the Bayesian brain hypothesis. After a description of the urn-ball task and methods for the acquisition of the ERP amplitudes for all four ERPs, the new Bayesian observer model is explained in detail before the DIF model is adapted to the urn-ball task. The effect of probability weighting on probabilistic reasoning is investigated by optionally incorporating weighting functions into the observer models. Next, the parameter optimization schemes as well as the composition of the design matrices for model estimation and selection (see Chap. 2) are specified. Results and conclusions complete this chapter, which was adapted and extended from Kolossa et al. (2015).

## 4.1 Overview

In the following, the urn-ball task and the Bayesian observer model are briefly described. The setup of the urn-ball task consists of different types of urns $u$, which are distinguished by the distribution of differently colored types of balls $k$ within them. Both the distribution of the urns and the distribution of the balls within these urns are made known to the subjects, thus giving them prior knowledge about the probability distributions over the urns and over the balls within the urns. One of the urns is then randomly drawn to form the *hidden* state $q(n) = u$, which is *not* revealed to the subjects during the experiment. From this urn, balls are sequentially drawn and take the form of observations $o(n) = k$ since they are shown to the subjects. The subjects have to respond to the ball color on each trial $n$ by pressing the respective

© Springer International Publishing Switzerland 2016

A. Kolossa, *Computational Modeling of Neural Activities for Statistical Inference*, DOI 10.1007/978-3-319-32285-8_4

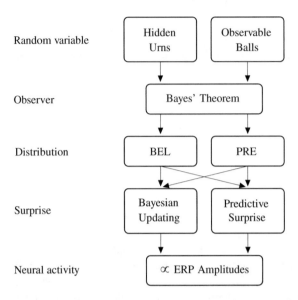

**Fig. 4.1** Hierarchical structure that relates random variables to probability distributions based on the Bayesian observer model. There are two random variables: Hidden states $q$ (selected urn type $u$) and observations $o$ (balls $k$ drawn). The Bayesian observer updates the probability distributions over hidden states (belief distribution, BEL) and observations (prediction distribution, PRE) following Bayes' theorem. Bayesian updating and predictive surprise are response functions that link the probability distributions to neural activities

key on a keyboard and they choose which urn they believe forms the hidden state after observing four balls.

This chapter proposes a Bayesian observer model which keeps track of two probability distributions, specifically the belief distribution over the urns, which is called BEL, and the prediction distribution over the observable balls, which is called PRE. Figure 4.1 shows the hierarchical structure which relates the random variables present in the urn-ball task to the ERP amplitudes via the Bayesian observer model with different kinds of surprise as response functions (see Sect. 1.4). There are two random variables: Events, i.e., balls which are sampled, and form observations $o(n) = k$, and the hidden states $q(n) = u$, i.e., the type of urn from which the balls are sampled. Bayesian updating and predictive surprise are used as response functions that link the BEL and PRE distributions (which the Bayesian observer model encodes) to measurable scalp potentials.

The potential role of nonlinear probability weighting (see Sect. 1.5) during probabilistic inference is also examined by applying (inverse) S-shaped probability weighting functions either to the inference input, yielding the $BEL_{SI}$ and $PRE_{SI}$ distributions, or to the inference output, yielding the $BEL_{SO}$ and $PRE_{SO}$ distributions, respectively. Table 4.1 presents all distributions that are based on the Bayesian observer model.

**Table 4.1** Overview and short description of the distributions based on the Bayesian observer model and ways to implement probability weighting

| Distribution | Description |
|---|---|
| BEL | Belief distribution about hidden states based on a Bayesian observer model |
| $BEL_{SI}$ | BEL distribution based on a Bayesian observer model employing S-shaped probability weighting of the inference input |
| $BEL_{SO}$ | BEL distribution based on a Bayesian observer model employing S-shaped probability weighting of the inference output |
| PRE | Prediction distribution about observations based on a Bayesian observer model |
| $PRE_{SI}$ | PRE distribution based on a Bayesian observer model employing S-shaped probability weighting of the inference input |
| $PRE_{SO}$ | PRE distribution based on a Bayesian observer model employing S-shaped probability weighting of the inference output |

The trial-by-trial ERP amplitudes were derived from the EEG signals and used for Bayesian model estimation and selection (see Chap. 2 for details). The analyses encompass the complete late positive complex that is known to be decomposable into three separable ERP components (Sutton and Ruchkin 1984; Dien et al. 2004): (1) The anteriorly distributed P3a component that usually occurs at latencies around 340 ms post-stimulus. (2) The parietally distributed P3b component at, in comparison to the P3a latency, variably delayed latencies. The functional significance of these two ERP components seems to be related to uncertainty (Sutton et al. 1965; Kopp and Lange 2013), surprise (Donchin 1981; Kolossa et al. 2013), decision-making (O'Connell et al. 2012; Kelly and O'Connell 2013), and Bayesian inference (Friston 2005; Kopp 2008). (3) A posteriorly distributed Slow Wave (SW) whose functional significance is, however, comparably less well understood (see Ruchkin et al. 1988; García-Larrea and Cézanne-Bert 1998; Spencer et al. 2001; Matsuda and Nittono 2015). The fronto-centrally distributed N250 is additionally analyzed as it is reported to be associated with decision making (Towey et al. 1980), as well as lower levels of attention and stimulus uncertainty (Strauss et al. 2015).

The digital filtering model, which showed clear superiority for the P300 amplitude fluctuations in Chap. 3, will be tested again against the Bayesian observer model. To this end, the model space contains postdictive and predictive surprise based on the DIF model with a uniform initial prior (DIF), with an objective initial prior ($DIF_{OP}$), and with an (inverse) S-shaped weighted initial prior ($DIF_{SP}$), respectively.

## 4.2 Participants, Experimental Design, Data Acquisition, and Data Analysis

### Participants

Sixteen undergraduate psychology students participated to gain course credits (fifteen women, one man, mean age: 24.7 years; age range 19–50 years). Handedness was examined with the Edinburgh Handedness Inventory (Oldfield 1971), revealing that one subject was left-handed and two were ambidextrous. All subjects indicated having normal or corrected-to-normal sight. The procedure was approved by the local Ethics Committee.

**Experimental Design**

The urn-ball task represents a modification of tasks that were used by Phillips and Edwards (1966) and Grether (1980, 1992) (see also Furl and Averbeck (2011), and Achtziger et al. (2014) for a similar task and FitzGerald et al. (2015) for a theoretical approach). There were $U = 2$ types of urns (labeled $u = 1$ and $u = 2$), which could be distinguished by the distribution of the $K = 2$ types of colored balls (labeled $k = 1$ and $k = 2$ for red and blue, respectively) of the ten balls contained in one urn. The urn types represented so-called states $q = u \in \mathcal{U} = \{1, 2\}$, which were hidden from the subjects during the experiment. The balls were so-called events $k \in \mathcal{K} = \{1, 2\}$, which could be observed by the subjects during the experiment. The distributions over the types of urns will be referred to as prior probabilities, and the distributions over the types of balls *within* the urns as likelihoods, respectively. The experimental design consisted of a factorial combination of two levels of prior probabilities (Pc (certain prior) and Pu (uncertain prior)) and two levels of likelihoods (Lc (certain likelihoods) and Lu (uncertain likelihoods)), yielding four experimental conditions $c \in \mathcal{C} = \{\text{PcLc, PcLu, PuLc, PuLu}\}$, each of which contained $B = 50$ episodes of sampling $b \in \{1, \ldots, 50\}$, each consisting of $N = 4$ trials $n \in \{1, \ldots, 4\}$, yielding a total of 800 sequentially presented colored ball stimuli. All conditions were administered to each of the subjects, with short breaks (approximately three minutes) between the conditions, and their order was counterbalanced across subjects. The ball colors were also counterbalanced across subjects, but this will be ignored in the description for better readability. As in the rest of this work, (●) represents the frequent event $k = 1$ and (●) the rare event $k = 2$.

Prior probabilities were manipulated by presenting ten urns, composed of different numbers of type $u = 1$ and type $u = 2$ urns. In uncertain prior probability conditions (Pu), seven type $u = 1$ urns and three type $u = 2$ urns (i.e., $P(q = 1) = 0.7$, $P(q = 2) = 0.3$) were presented. In the certain prior probability condition (Pc), nine type $u = 1$ urns and one type $u = 2$ urn (i.e., $P(q = 1) = 0.9$, $P(q = 2) = 0.1$) were presented. On uncertain likelihood conditions (Lu), urn type $u = 1$ contained seven red ($k = 1$) and three blue ($k = 2$) balls (i.e., $P(o = 1|q = 1) = 0.7$, $P(o = 2|q = 1) = 0.3$), while urn type $u = 2$ contained three red and seven blue balls (i.e., $P(o = 1|q = 2) = 0.3$, $P(o = 2|q = 2) = 0.7$). On certain likelihood conditions (Lc), urn type $u = 1$ contained nine red balls and one blue ball (i.e., $P(o = 1|q = 1) = 0.9$, $P(o = 2|q = 1) = 0.1$), while urn type $u = 2$ contained one red ball and nine blue balls (i.e., $P(o = 1|q = 2) = 0.1$, $P(o = 2|q = 2) = 0.9$).

At the beginning of each condition, the *tableau* of ten urns containing a total of 100 balls was shown to the subjects, representing prior probabilities and likelihoods. The visualization of prior probabilities and likelihoods in the form of *tableaus* allowed the subjects to build an internal representation of these probabilistic parameters. Figure 4.2 illustrates the experiment for one episode of sampling in the PuLu

Condition:  $c =$ PuLu

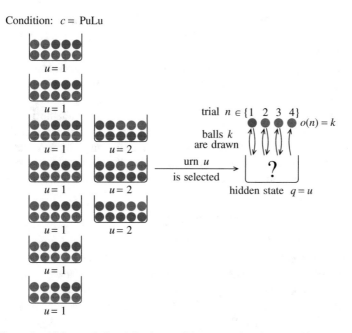

**Fig. 4.2** Illustration of the urn-ball task for the condition uncertain prior probability and uncertain likelihood PuLu, i.e., $P(q=1) = 0.7$, $P(q=2) = 0.3$, $P(o=1|q=1) = P(o=2|q=2) = 0.7$, and $P(o=2|q=1) = P(o=1|q=2) = 0.3$. At the beginning of a condition, the tableau of *urns u* and *balls k* is shown to the participant, describing prior probability and likelihood. Afterwards, an urn is randomly selected to constitute the hidden state $q=u$ and not shown to the participant. From this urn, *four balls* are drawn consecutively over trials $n \in \{1, 2, 3, 4\}$ with replacement, and observed by the subjects $o(n)=k \in \{1, 2\}$. The subjects are asked to indicate the color of each ball stimulus by pressing the corresponding key and have to choose which type of urn had been selected on the current episode of sampling once a sample of *four balls* is completed

condition. Each episode of random sampling consisted of the following sequence of events: First, one of the ten urns was selected randomly to form the state $q=u$, but the outcome of this selection remained hidden to the subjects during the experiment. Subsequently, a random sample of four balls was sequentially drawn, with replacement, from the selected urn, and shown one after the other, taking the form of observations $o(n)=k$ at each trial $n$.

The subjects were asked to indicate the color of each ball stimulus by pressing the left or right Ctrl key on a standard computer keyboard (using the left or right index finger, respectively). Once a sample of four balls had been completed, the subjects had to choose which type of urn had been selected on the current episode of sampling (i.e., which urn type $u$ constitutes the state $q$). They indicated their choice by pressing the left or right Ctrl key for the state being urn type $q=u=1$ and urn type $q=u=2$, respectively. Stimulus-response mapping was counterbalanced across subjects (i.e., left- or right-hand responses indicating $q=1$ and $q=2$ choices). The duration of an episode of sampling, i.e., the presentation of the sample of ball stimuli, the collection of the color-specific responses, and the final urn choice, amounted to around 12 to 15 seconds. Neither feedback nor reward was provided during the course of the experiment.

Before an experiment was actually carried out, each subject completed four practice episodes of sampling under the supervision of the experimenter to become accustomed to the task. The tableau of each practice sampling consisted of one single $u=1$ urn and one single $u=2$ urn (yielding uniform prior probabilities). Successful completion of these practice episodes of samplings demonstrated that the subjects understood the procedure and their task. Visual ball stimuli were presented at each trial for a duration of 100 ms each, with a stimulus-to-stimulus interval of 2.5 s. Stimuli were displayed at the center of the monitor against a light gray background with a viewing distance of 1.25 m.

**Selection of ERP Data for Further Analyses**

Event-related potential (ERP) waveforms were created for each subject $\ell$, electrode e, and experimental condition $c$ by dividing the continuous EEG into epochs of 700 ms duration, starting 100 ms before stimulus onset, yielding $y_{n,\ell,c,b,e}(t)$, with block index $b$ and trial index $n$ (see (1.1) in Sect. 1.1 for details). Epochs were corrected using the interval $[-100, 0]$ ms before stimulus presentation as the baseline (1.2). For each of the four ERP components, (N250, P3a, P3b, SW) a specific region of interest was defined based on the literature. The resulting electrode sets equaled $\mathcal{E}_{N250} = \{C3, Cz, C4\}$ for the N250 (Towey et al. 1980), $\mathcal{E}_{P3a} = \{Fz, FCz, Cz\}$ for the P3a (Kopp and Lange, 2013), $\mathcal{E}_{P3b} = \{Pz\}$ for the P3b (Kolossa et al. 2013), and $\mathcal{E}_{SW} = \{O1, O2\}$ for the SW (Matsuda and Nittono 2015). A single virtual electrode was then created for each ERP $\in \{N250, P3a, P3b, SW\}$ by averaging the epochs over the entire set of respective electrodes $\mathcal{E}_{ERP}$, i.e.,

$$y_{n,\ell,c,b,\mathrm{ERP}}(t) = \frac{1}{|\mathcal{E}_{\mathrm{ERP}}|} \sum_{e \in \mathcal{E}_{\mathrm{ERP}}} y_{n,\ell,c,b,e}(t), \tag{4.1}$$

with $|\mathcal{E}_{ERP}|$ being the number of electrodes in $\mathcal{E}_{ERP}$. Event-related potential (ERP) waveforms $\overline{y}_{\ell,c,k,\mathrm{ERP}}(t)$ (compare to (1.3)) were then created by extracting a single epoch per trial $n$ over blocks $b$, separating them by event type $k$, and subsequent averaging over the epochs (see Sect. 1.1 for details on epoch extraction and ERP waveform calculation). These ERP waveforms were then averaged over the subjects to obtain the grand-average ERP waveforms:

$$\overline{y}_{c,k,\mathrm{ERP}}(t) = \frac{1}{L} \sum_{\ell=1}^{L} \overline{y}_{\ell,c,k,\mathrm{ERP}}(t). \tag{4.2}$$

The ERP component latencies $t_{\mathrm{ERP}}$ were determined through the analysis of grand-average ERP waveform variability over event types and experimental conditions at these virtual electrodes. Specifically, the grand-average ERP waveforms for frequent and rare events in each of the four experimental conditions were analyzed with regard to the latencies at which maximum variability between ERP waveforms occurred

$$t_{\mathrm{ERP}} = \underset{t \in \mathcal{T}_{\mathrm{ERP}}}{\arg\max} \left\{ \frac{1}{|\mathcal{C}| \cdot K - 1} \sum_{c \in \mathcal{C}} \sum_{k \in \mathcal{K}} \left( \overline{y}_{c,k,\mathrm{ERP}}(t) - \overline{y}_{\mathrm{ERP}}(t) \right)^2 \right\}, \qquad (4.3)$$

with

$$\overline{y}_{\mathrm{ERP}}(t) = \frac{1}{|\mathcal{C}| \cdot K} \sum_{c \in \mathcal{C}} \sum_{k \in \mathcal{K}} \overline{y}_{c,k,\mathrm{ERP}}(t), \qquad (4.4)$$

and $|\mathcal{C}|$ denoting the number of conditions in the set $\mathcal{C}$. Maximum variance was searched within the predefined time intervals of $\mathcal{T}_{\mathrm{N250}} = [200, 300]\,\mathrm{ms}$ for the N250 (Towey et al. 1980), of $\mathcal{T}_{\mathrm{P3a}} = \mathcal{T}_{\mathrm{P3b}} = [300, 400]\,\mathrm{ms}$ for the P3a and P3b (Kolossa et al. 2013; Kopp and Lange 2013), and of $\mathcal{T}_{\mathrm{SW}} = [400, 560]\,\mathrm{ms}$ for the SW (García-Larrea and Cézanne-Bert 1998). The thus derived values for $t_{\mathrm{ERP}}$ were $t_{\mathrm{N250}} = 232\,\mathrm{ms}$, $t_{\mathrm{P3a}} = 356\,\mathrm{ms}$, $t_{\mathrm{P3b}} = 380\,\mathrm{ms}$, and $t_{\mathrm{SW}} = 504\,\mathrm{ms}$. Single-trial ERP amplitudes $y_{\ell,c,b,\mathrm{ERP}}(n)$ were finally obtained by taking the smoothed values (compare to (1.5)) at time point $t_{\mathrm{ERP}}$ from the single-epoch EEG signal $y_{n,\ell,c,b,\mathrm{ERP}}(t)$ with the smoothing interval of $t_{\mathrm{ERP}} \pm 20\,\mathrm{ms}$ (see Sect. 1.1 for details on the acquisition of the single-trial ERP amplitudes). The interval width was chosen smaller than in Chap. 3 to provide smoothed amplitudes at the moment of peak variance but at the same time to enable a dissociation of the late positive complex. Specifically, the P3a and P3b are in close temporal proximity (Polich 2007; Kopp 2008) and a greater smoothing interval width might increase interference between the two ERPs via volume conduction (Luck 2014).

## 4.3   The Bayesian Observer Model

In the following, the Bayesian observer model is deduced from probability theory. The belief distribution (BEL) and prediction distribution (PRE) are then derived from the Bayesian observer, and optional hyper-parameterization using probability weighting functions is described. The application of the response functions precedes a graphical summary of Bayesian inference.

### 4.3.1   Bayes' Theorem and the Urn-Ball Task

Let $\mathrm{P}(q(n), \mathbf{o}_1^n)$ be the joint probability of state $q(n)$, and the sequence of previous $\mathbf{o}_1^n$. The sequence $\mathbf{o}_1^n$ can be decomposed into the sequence of previous observations $\mathbf{o}_1^{n-1}$ and the current observation $o(n)$, which yields

$$\mathrm{P}(q(n), \mathbf{o}_1^n) = \mathrm{P}(q(n), \underbrace{\mathbf{o}_1^{n-1}, o(n)}_{\mathbf{o}_1^n}). \qquad (4.5)$$

Additionally, the joint probability $P(q, \mathbf{o}_1^n)$ can be expanded using the chain rule following

$$P(q(n), \mathbf{o}_1^n) = P(q(n)|\mathbf{o}_1^n) \cdot P(\mathbf{o}_1^n), \tag{4.6}$$

or, equivalently,

$$P(q(n), \mathbf{o}_1^{n-1}, o(n)) = P(o(n)|q(n), \mathbf{o}_1^{n-1}) \cdot P(q(n), \mathbf{o}_1^{n-1}) = P(q(n), \mathbf{o}_1^n). \tag{4.7}$$

Next, solving (4.7) for $P(q(n)|\mathbf{o}_1^n)$ yields

$$P(q(n)|\mathbf{o}_1^n) = \frac{P(o(n)|q(n), \mathbf{o}_1^{n-1}) \cdot P(q(n), \mathbf{o}_1^{n-1})}{P(\mathbf{o}_1^n)} \tag{4.8}$$

$$= \frac{P(o(n)|q(n), \mathbf{o}_1^{n-1}) \cdot P(q(n)|\mathbf{o}_1^{n-1}) \cdot P(\mathbf{o}_1^{n-1})}{P(o(n)|\mathbf{o}_1^{n-1}) \cdot P(\mathbf{o}_1^{n-1})} \tag{4.9}$$

$$= \frac{P(o(n)|q(n), \mathbf{o}_1^{n-1}) \cdot P(q(n)|\mathbf{o}_1^{n-1})}{P(o(n)|\mathbf{o}_1^{n-1})}. \tag{4.10}$$

Since the urn-ball task is designed with replacement, knowing the urn type, earlier observations do not influence the probability of the current observation

$$P(o(n)|q(n), \mathbf{o}_1^{n-1}) \approx P(o(n)|q(n)). \tag{4.11}$$

After inserting (4.11) in (4.10), Bayes' theorem is derived in the form

$$P(q(n)|\mathbf{o}_1^n) = \frac{P(o(n)|q(n)) \cdot P(q(n)|\mathbf{o}_1^{n-1})}{P(o(n)|\mathbf{o}_1^{n-1})}. \tag{4.12}$$

The Bayesian observer model simply follows Bayes' theorem (4.12). The belief about the hidden state $q(n)$ being urn type $u$ before observing $o(n) = k$ is called the predictive prior probability $P(q(n) = u|\mathbf{o}_1^{n-1}) = P_u(n-1)$, while after observing $o(n) = k$ it becomes the posterior probability $P(q(n) = u|\mathbf{o}_1^n) = P_u(n)$. Note that the predictive prior probability is simply referred to as the prior probability in this work to avoid confusion with the fundamentally different prediction distribution. The distribution of balls within a certain urn equals the likelihood $P(o(n) = k|q(n) = u) = \mathcal{L}_{k|u}$, and the prediction of the observation constitutes the observation probability $P(o(n) = k|\mathbf{o}_1^{n-1}) = P_k(n)$. The prior $P(q(n) = u|\mathbf{o}_1^{n-1})$ on trial $n$ is calculated based on the preceding posterior $P(q(n-1) = u|\mathbf{o}_1^{n-1})$ on trial $n-1$ following

$$P(q(n) = u|\mathbf{o}_1^{n-1}) = \sum_{u' \in \mathcal{U}} P(q(n) = u|q(n-1) = u') \cdot P(q(n-1) = u'|\mathbf{o}_1^{n-1}), \tag{4.13}$$

with the state transition probability

$$P(q(n)=u|q(n-1)=u') = \begin{cases} 1, & \text{if } u=u' \\ 0, & \text{otherwise,} \end{cases} \tag{4.14}$$

implying that within one episode of sampling the chosen urn does not change. The state transition probabilities (4.14) can be equivalently expressed in a more intuitive fashion as a state transition matrix, shown here for $\mathcal{U}=\{1,2\}$:

$$\begin{bmatrix} P(q(n)=1|q(n-1)=1) & P(q(n)=2|q(n-1)=1) \\ P(q(n)=1|q(n-1)=2) & P(q(n)=2|q(n-1)=2) \end{bmatrix} = \begin{bmatrix} 1 & 0 \\ 0 & 1 \end{bmatrix}. \tag{4.15}$$

The elements on the main diagonal represent the probability for the state to remain the same, while the other elements represent the probability for a state switch. Note that if the hidden state could change from one trial to the next, the elements on the main diagonal would be smaller than one, while the other elements would be larger than zero.

The calculation of posterior beliefs $P_{\mathcal{U}}(n)$ is fundamental to the BEL distribution, while the PRE distribution rests upon the calculation of the prediction distribution $P_{\mathcal{K}}(n+1)$, which in turn is based on these posterior beliefs.

## 4.3.2 The Belief Distribution (BEL)

The belief distribution (BEL) $P(q(n) - u|\mathbf{o}_1^n)$, $\forall u \in \mathcal{U}$ is the posterior distribution of the hidden state $q(n)$ being urn type $u$ based on Bayes' theorem (4.12). The likelihood term $P(o(n) = k|q(n) = u) = \mathcal{L}_{k|u}$ is defined as the probability of the observation $o(n)=k$, given that the events are originating from state $q(n)=u$, and the prior $P(q(n) = u|\mathbf{o}_1^{n-1}) = P_u(n-1)$ as the probability of state $q(n) = u$, given a sequence $\mathbf{o}_1^{n-1} = (o(1), o(2), \ldots, o(n-1))$ of $n-1$ previous observations. The initial prior probability $P(q(n=1)=u)=P_u$ and likelihoods were described to the subjects at the beginning of each experimental condition by presenting the tableau of all urns with their respective ball distributions. Note that after (4.11) the likelihood term remains constant throughout all trials $n \in \{1, \ldots, 4\}$. The posterior probability $P(q(n) = u|\mathbf{o}_1^n) = P_u(n)$ after observation $o(n)$ is evaluated according to Bayes' theorem (4.12)

$$P_u(n) = \frac{\mathcal{L}_{k|u}P_u(n-1)}{P_k(n)}, \tag{4.16}$$

with the observation probability

$$P_k(n) = P(o(n)=k|\mathbf{o}_1^{n-1}) = \sum_{u \in \mathcal{U}} P(o(n)=k|q(n)=u) \cdot P(q(n)=u|\mathbf{o}_1^{n-1}) \quad (4.17)$$

$$= \sum_{u \in \mathcal{U}} \mathcal{L}_{k|u} P_u(n-1).$$

Calculating (4.16) for all $u \in \mathcal{U} = \{1, \ldots, U\}$ yields the posterior distribution $P_{\mathcal{U}}(n)$.

### 4.3.3 The Prediction Distribution (PRE)

The prediction distribution (PRE) $P(o(n+1)=k|\mathbf{o}_1^n)$, $k \in \mathcal{K}$ estimates the probability distribution over future observations based on the posterior distribution $P_{\mathcal{U}}(n)$ by calculating

$$P_k(n+1) = P(o(n+1)=k|\mathbf{o}_1^n) = \sum_{u \in \mathcal{U}} P(o(n+1)=k|q(n+1)=u) \cdot P(q(n)=u|\mathbf{o}_1^n)$$

$$(4.18)$$

for all $k \in \mathcal{K} = \{1, \ldots, K\}$. As the urn does not change within an episode of sampling, it follows $P(o(n+1) = k|q(n+1) = u) = P(o(n) = k|q(n) = u)$, and (4.18) can be simplified to

$$P_k(n+1) = \sum_{u \in \mathcal{U}} \mathcal{L}_{k|u} P_u(n). \quad (4.19)$$

### 4.3.4 Surprise Based on the Bayesian Observer Model

This section shows how Bayesian surprise, postdictive surprise, and predictive surprise are applied as response functions to the BEL and PRE distributions. See Sect. 1.4 for details on the different kinds of surprise.

#### 4.3.4.1 Surprise Based on the BEL Distribution

**Bayesian Surprise**

Bayesian surprise $I_B$ (1.19) is obtained as the Kullback–Leibler divergence $D_{KL}$ between the prior distribution $P_{\mathcal{U}}(n-1)$ and the posterior distribution $P_{\mathcal{U}}(n)$

$$I_B(n) = D_{KL}(P_{\mathcal{U}}(n-1) \| P_{\mathcal{U}}(n)). \quad (4.20)$$

Notice that Bayesian surprise reflects the degree of Bayesian updating, because the prior distribution on any trial equals the posterior distribution on the previous trial. In short, Bayesian surprise reflects the changes in beliefs over hidden states that are induced by observations.

**Predictive Surprise**

As detailed in Sect. 4.2, the hidden state is not revealed to the subjects at any time. It is therefore not feasible to calculate Shannon surprise $I_P(n) = -\log_2 P_u(n)$ in relation to the state. However, the entropy $I_H(n)$ (1.21) of the belief distribution, which is the average Shannon surprise, is used as response function (see Sect. 1.4.2):

$$I_H(n) = -\sum_{u \in \mathcal{U}} P_u(n) \log_2 P_u(n). \tag{4.21}$$

### 4.3.4.2 Surprise Based on the PRE Distribution

**Postdictive Surprise**

Postdictive surprise $I_B(n)$ (1.18) is defined analogously to Bayesian surprise (4.20) via the prediction distributions before an observation $P_{\mathcal{K}}(n)$ and after an observation $P_{\mathcal{K}}(n+1)$:

$$I_B(n) = D_{KL}(P_{\mathcal{K}}(n) \| P_{\mathcal{K}}(n+1)). \tag{4.22}$$

Postdictive surprise reflects the degree of Bayesian updating as it quantifies the changes in predictions about observable events that are induced by observations.

**Predictive Surprise**

In contrast to postdictive surprise, predictive surprise $I_P(n)$ (1.20) is the surprise about the current observation $o(n)$ at trial $n$ being $k$ under the prediction $P_k(n) \in P_{\mathcal{K}}(n)$ after a sequence $\mathbf{o}_1^{n-1} = (o(1), o(2), \dots, o(n-1))$ of $n-1$ former observations:

$$I_P(n) = -\log_2 P_k(n). \tag{4.23}$$

Note that $P_k(n)$ is the denominator of (4.16), i.e., the one probability taken from the prediction distribution $P_{\mathcal{K}}(n)$ that corresponds to the actual observation $o(n)$.

## 4.3.5 Summary and Visualization of Bayesian Inference

Bayesian inference is visualized in Fig. 4.3: It illustrates likelihoods, beliefs, predictions, and their trial-by-trial updating by an ideal Bayesian observer in the PuLu condition. The left panel shows the probability distributions and likelihoods on trial $n-1$, while the right panel shows these quantities on trial $n$ after a blue colored

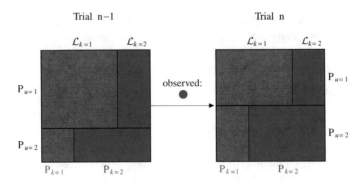

**Fig. 4.3** Illustration of Bayesian inference in the PuLu condition: Updating of the belief distribution $P_\mathcal{U}$ and of the prediction distribution $P_\mathcal{K}$ under uncertain prior probabilities and likelihoods $\mathcal{L}_k$. *Left panel* Likelihoods and probability distributions on trial $n-1$. *Right panel* Likelihoods and probability distributions on trial $n$. The *horizontal lines* define the probability distribution over the *hidden state* (beliefs about urns), while *vertical lines* define the *likelihoods* (ball distribution within the urns). *Colored areas* visualize the resulting *prediction distribution* for *red balls* (▮) versus *blue balls* (▮), respectively. The *arrow* indicates the observation of a *blue ball* (●) on trial $n$ ($o(n) = 2$). The observation of the *blue ball* triggers Bayesian inference about the hidden state that is equivalent to shifting the *horizontal line*, and the resulting ratio of colored areas on trial $n$ represents the updated prediction distribution

ball (●) was observed ($o(n)=2$). The horizontal line defines the beliefs ($P_{u=1}$, $P_{u=2}$), i.e., the probability distribution over the hidden states, while the vertical lines define the likelihoods ($\mathcal{L}_{k=1}$, $\mathcal{L}_{k=2}$), i.e., the distribution of balls within each type of urn which did not change within conditions. The colored areas visualize the resulting prediction distribution over events ($P_{k=1}$ (▮), $P_{k=2}$ (▮)).

The arrow indicates the presentation of a (surprising, since rare) blue ball (($●$), $o(n) = 2$) which triggers predictive surprise (4.23) and Bayesian inference (BEL distribution, $P_\mathcal{U}(n-1) \to P_\mathcal{U}(n)$), equaling shifts of the horizontal line. In this case, predictive surprise about the blue ball exceeds predictive surprise about a (potential) red ball since $P_{k=2}(n) < P_{k=1}(n)$. The resulting ratio of colored areas represents the new prediction distribution ($P_\mathcal{K}(n) \to P_\mathcal{K}(n+1)$). The KL divergence (4.20) between $P_\mathcal{U}(n-1)$ and $P_\mathcal{U}(n)$ yields the scalar values used to predict trial-by-trial ERP amplitudes based on the updating of the BEL distribution (Bayesian surprise). The KL divergence (4.22) between $P_\mathcal{K}(n)$ and $P_\mathcal{K}(n+1)$ yields the scalar values used to predict trial-by-trial ERP amplitudes based on the updating of the PRE distribution (postdictive surprise).

## 4.4 Incorporating Probability Weighting Functions into the Bayesian Observer Model

This section describes two ways of how probability weighting functions as introduced in Sect. 1.5 can be incorporated into the Bayesian observer model and how the shape parameter $\zeta$ is estimated.

### 4.4.1 Probability Weighting of the Inference Input (BEL$_{SI}$ and PRE$_{SI}$)

Probability weighting can be incorporated into the Bayesian observer model as a hyper-parameterization *of all input* of the inference (i.e., prior probabilities and likelihoods). Bayesian inference takes place as before, yielding the BEL$_{SI}$ distribution (4.16)

$$P_u(n) = \frac{\mathcal{L}_{k|u}^{(w)} P_u^{(w)}(n-1)}{\sum_{u \in \mathcal{U}} \mathcal{L}_{k|u}^{(w)} P_u^{(w)}(n-1)}, \quad \forall u \in \mathcal{U}, \tag{4.24}$$

and the PRE$_{SI}$ distribution (4.19)

$$P_k(n+1) = \sum_{u \in \mathcal{U}} \mathcal{L}_{k|u}^{(w)} P_u^{(w)}(n), \quad \forall k \in \mathcal{K}. \tag{4.25}$$

Note that the posterior probability from the preceding trial is weighted when it becomes the prior probability for the current trial. This is a consequence of the hyper-parameterization of *all* probabilities. Figure 4.4 is a block diagram of the Bayesian observer model which further illustrates the weighting of the inference input. The initial prior $P_u$ and likelihood $\mathcal{L}_{k|u}$ were described to the subjects at the beginning of each experimental condition by a view of the tableau with all possible urns (initial prior probabilities) with their respective ball distributions (likelihoods). The described probabilities are being presented via dashed lines and the experience via solid lines. The transition from probability description to actual experience of the probabilities is illustrated by the switch which turns at $n = 2$. On the first trial, the initial prior probabilities and likelihoods are weighted, while on the following trials the posterior probabilities are weighted when becoming the new prior probabilities. See the feedback path with the trial delay element $\boxed{T}$ for a graphical representation of the transition from posterior probability to prior probability, and Sect. 4.3.1 for a formal derivation. Note that the application of Bayesian updating and predictive surprise as response functions is not altered by probability weighting and takes place as described in Sect. 4.3.4.

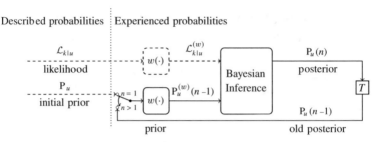

**Fig. 4.4** Block diagram of the Bayesian observer with hyper-parameterization of the inference input. At the beginning of an experimental condition, the initial prior probabilities $P_u$ and likelihoods $\mathcal{L}_{k|u}$ are described to the subjects, which can be interpreted as the initialization of the model (*dashed part* of the diagram). At trial $n = 2$ the switch is turned to close the feedback path with the trial delay element $\boxed{T}$, so the posterior $P_u(n-1)$ of trial $n-1$ is used as prior probability on trial $n$. All probabilities are weighted (see Sect. 1.5) *before being input* into the Bayesian inference process (4.24), yielding the posterior probability $P_u(n)$ of trial $n$

Empirically derived values of the weighting parameters are of great variability (Cavagnaro et al. 2013). Zhang and Maloney (2012) propose $P_0 = \frac{1}{K} = \frac{1}{U} = 0.5$ which, according to (1.25), ensures that the same weighting functions can be applied to all probabilities while meeting the probabilistic constraints $\sum_{k \in \mathcal{K}} \mathcal{L}_{k|u}^{(w)} = 1$ and $\sum_{u \in \mathcal{U}} P_u^{(w)}(n-1) = 1$. This keeps the complexity of the observer model as low as possible as only the single-shape parameter $\zeta$ is a free parameter. The optimization of this free parameter will be shown later on in Sect. 4.4.3. Figure 4.5 illustrates the influence of probability weighting, constrained in the manner described above, on Bayesian inference. The effect of probability over- and underestimation clearly persists on posterior probabilities that are biased towards $P(q = 1|o = k) = 0.5$.

### 4.4.2  Probability Weighting of the Inference Output (BEL$_{SO}$ and PRE$_{SO}$)

Probability weighting for the BEL$_{SI}$ and PRE$_{SI}$ distributions equals a processing step which precedes inference, because the input is weighted before Bayes' theorem is applied (i.e., all probabilities and likelihoods on the right-hand side of (4.16)). Alternatively, probability weighting could be implemented as a processing step which succeeds inference. This means that the output (i.e., the BEL and PRE distributions themselves) is weighted before response functions (see Sect. 4.3.4) are applied. Following (4.16) and (4.19) yields the BEL$_{SO}$ distribution

$$P_u^{(w)}(n) = w(P_u(n)), \quad \forall u \in \mathcal{U}, \tag{4.26}$$

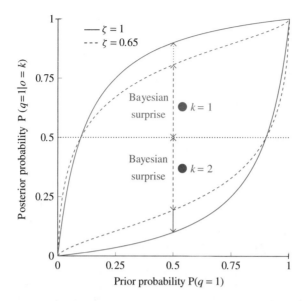

**Fig. 4.5** Comparison of posterior probabilities $P(q = 1|o = k)$ for frequent events ($(\bullet)$, $k = 1$, (——) and (- - -)), and rare events ($(\bullet)$, $k = 2$, (——) and (- - -)), calculated via Bayes' theorem with unweighted prior probability $P(q = 1)$ and exemplary likelihood $P(o = 1|q = 1) = 0.9$ ((4.16), (——) and (——)), and weighted prior probability and likelihood ((4.24), with inverse S-shaped weighting (1.22) with $\zeta = 0.65$, (- - -) and (- - -)). (·····) represents a posterior probability of 0.5. For both frequent and rare events, weighting leads to a bias towards higher uncertainties in posterior probabilities. The *double-headed arrows* illustrate the quantity of Bayesian surprise (4.20) for a prior probability of $P(q = 1) = 0.5$ for a frequent event (($\leftarrow$- -$\rightarrow$) with weighting, ($\leftarrow$- -$\rightarrow$ + ——) without weighting), and a rare event (($\leftarrow$- -$\rightarrow$) with weighting, ($\leftarrow$- -$\rightarrow$ + ——) without weighting)

and the PRE$_{SO}$ distribution

$$P_k^{(w)}(n+1) = w(P_k(n+1)), \quad \forall k \in \mathcal{K}, \tag{4.27}$$

respectively. This can be interpreted as a nonlinearity which exists only between the non-weighted probability distributions and the response functions rather than a proper nonlinear weighting of probabilities.

### 4.4.3 Weighting Parameter Optimization

In contrast to the parameters of the DIF model (see Sect. 3.4.6), the hyper-parameter $\zeta$ cannot be optimized based on the EEG data, as only $N = 200$ trials were recorded in each experimental condition for the urn-ball task. Splitting the data into a test data set and a training data set would not leave enough trials in either set for reliable model selection results (see Sect. 2.5). Furthermore, four distinct ERPs (N250, P3a, P3b,

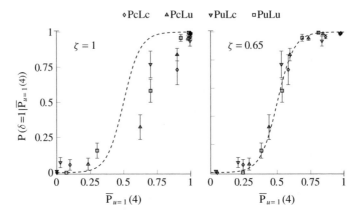

**Fig. 4.6** Likelihoods (ratios of choices) $P(\delta = 1|\overline{P}_{u=1}(4))$ of choosing urn type $u = 1$ based on all subjects in dependence on the mean posterior probability $\overline{P}_{u=1}(4)$ (4.16) of the BEL (*left panel*) and $BEL_{SI}$ (*right panel*) distributions after observing an episode of sampling. Average posterior probabilities were calculated over all episodes containing identical ratios of types of balls. *Error bars* indicate standard error between subjects. The hyper-parameter $\zeta$ was optimized by minimizing the mean squared error (MSE) between the optimal choice function indicated by the *dashed black line* and the measured likelihoods resulting in $\zeta = 0.65$

and SW) define the regions of interest in the EEG data, and ERP-specific optimization would be counter-intuitive to a single Bayesian observer. For the specific setup of the urn-ball task, Zhang and Maloney (2012) propose $\zeta = \frac{\log 104}{\log 100} \approx 1.0$, which yields the unaltered BEL and PRE distributions. However, the behavioral data, i.e., the chosen urns at the end of each episode of sampling, can be used for the optimization of $\zeta$.

Figure 4.6 shows the likelihoods $P(\delta = 1|\overline{P}_{u=1}(4))$ based on all subjects for choice $\delta \in \mathcal{U}$ of the subjects being urn type $u = 1$, depending on the average posterior probability $\overline{P}_{u=1}(4)$ after observing an episode of sampling. The posterior probabilities were averaged over all sequences containing identical ratios of types of ball colors. Assuming that the decision depends on the posterior probabilities and that these probabilities are used optimally, the general form of the likelihood function in a binary decision task is sigmoid (Diaconescu et al. 2014) and of the form (Daunizeau et al. 2014)

$$P_{opt}(\delta = 1|\overline{P}_{u=1}(4)) = \frac{1}{1 + e^{-a(\overline{P}_{u=1}(4)-b)}}. \tag{4.28}$$

Values a and b control the link between choices and posterior probabilities and a bias, respectively. The parameter b = 0.5 is chosen because both choice options are equally preferred and thus no bias is to be expected. The parameter a is set to 16 yielding a steep slope around $\overline{P}_{u=1}(4) = 0.5$. The dashed black line indicates this optimal choice function. The hyper-parameter $\zeta$, which controls the curvature of the inverse S-shaped weighting function (1.22) for the $BEL_{SI}$ distribution, is optimized by minimizing the mean squared error (MSE) between the measured likelihoods $P(\delta = 1|\overline{P}_{u=1}(4))$ and the optimal likelihood function $P_{opt}(\delta = 1|\overline{P}_{u=1}(4))$:

$$\zeta_{opt} = \arg\min_{\zeta} \left\{ MSE\left(P(...), P_{opt}(...)\right) \right\}. \tag{4.29}$$

Evaluating the range $\zeta \in [0.5, \ldots, 1]$ with an increment of 0.01 yields $\zeta_{opt} = 0.65$. The left panel of Fig. 4.6 shows that for the BEL distribution (4.16), the data points are spread out around the optimal choice function, while the right panel shows that for the $BEL_{SI}$ distribution (4.24) with $\zeta = 0.65$ the data points are clustered around the optimal choice function. Note that since a and b were set based on plausible assumptions as described above, this is the single free parameter of the Bayesian observer model, which was optimized on the group level for all subjects using behavioral data. This parameter estimate is used for the $BEL_{SI}$, $BEL_{SO}$, $PRE_{SI}$, and $PRE_{SO}$ distributions. Note that using this procedure, the $PRE_{SI}$, $BEL_{SO}$, and $PRE_{SO}$ distributions are not explicitly optimized. Nevertheless, *if* the inference input is weighted, it is reasonable to assume identical weighting parameters for the $BEL_{SI}$ and $PRE_{SI}$ distributions. Furthermore, *if* the inference output is subject to probability weighting functions, i.e., if there is a the nonlinearity between the distributions and the response functions, any parameter $\zeta < 1$ should result in superiority of the $BEL_{SO}$ and $PRE_{SO}$ distributions, so $\zeta = 0.65$ can be used to test this hypothesis.

## 4.5   The DIF Model in the Urn-Ball Task

The digital filtering model is put in competition with the Bayesian observer model. To this end, the free model parameters are set to the values derived in Sect. 3.6.2.1. In order to adapt the DIF model to the urn-ball task, it is tested with a uniform initial prior (DIF), an objective initial prior ($DIF_{OP}$), and an (inverse) S-shaped weighted initial prior $DIF_{SP}$. Table 4.2 gives an overview on the versions of the DIF model which are tested in the urn-ball task. For all versions, postdictive surprise and predictive surprise are used as response functions as described in Sect. 3.4.5.

**Table 4.2** Overview and short description of the versions of the DIF model which are tested in the urn-ball task

| Model | Description |
| --- | --- |
| DIF | DIF model with a uniform initial prior as introduced in Chap. 3 |
| $DIF_{OP}$ | DIF model with the objective initial prior |
| $DIF_{SP}$ | DIF model with an (inverse) S-shaped weighted initial prior |

### 4.5.1  The Objective Initial Prior (DIF$_{OP}$)

For the DIF$_{OP}$ model, the objective initial prior is made known to the model by using the marginal event probability $P_k$ instead of a uniform initial prior probability in the input signal (3.19):

$$g_k(n) = \begin{cases} P_k, & \text{if } n \leq 0 \text{ (objective initial prior)} \\ 1, & \text{if } n > 0 \text{ and } o(n) = k \\ 0, & \text{otherwise,} \end{cases} \tag{4.30}$$

with

$$P_k = \sum_{u \in \mathcal{U}} \mathcal{L}_{k|u} P_u. \tag{4.31}$$

Note that (4.31) equals the PRE distribution (4.19) for $n=0$, with the likelihood $\mathcal{L}_{k|u}$ and the initial prior probability $P_u$.

### 4.5.2  The Initial Prior Using Weighting Functions (DIF$_{SP}$)

For the DIF$_{SP}$ model, the weighted initial prior is calculated by applying an (inverse) S-shaped probability weighting function (see Sect. 1.5). As for the PRE$_{SI}$ distribution (4.27), the likelihood $\mathcal{L}_{k|u}$ and initial prior probability $P_u$ are weighted before calculating the marginal ball distribution $P_k$:

$$P_k = \sum_{u \in \mathcal{U}} \mathcal{L}_{k|u}^{(w)} P_u^{(w)}, \tag{4.32}$$

which is then used as initial prior in the input signal (4.30):

$$g_k(n) = \begin{cases} P_k, & \text{if } n \leq 0 \text{ (weighted initial prior)} \\ 1, & \text{if } n > 0 \text{ and } o(n) = k \\ 0, & \text{otherwise.} \end{cases} \tag{4.33}$$

As stated in Sect. 4.4.3, an EEG data-driven optimization of $\zeta$ is not feasible. In order to test whether probability weighting improves the performance of the DIF model at all, $\zeta = 0.65$ is used for the DIF$_{SP}$ model, same as for the BEL$_{SO}$ and PRE$_{SO}$ distributions.

## 4.6 Specification of the Design Matrices for Model Estimation and Selection in the Urn-Ball Task

As in Chap. 3, the parametric empirical Bayes framework (see Sect. 2.4) is used for model estimation (see Sect. 2.4.2), and model selection rests upon posterior model probabilities for group studies $P(m|\mathbf{y})$ derived from the variational free energy $F_m$ (see Sect. 2.4.4). In contrast to Chap. 3, four different ERPs are analyzed, which is simply a repetition of the following procedure for each ERP $\in$ {N250, P3a, P3b, SW}. To keep the presentation simple, the corresponding subscript is dropped in the following. The model space $\mathcal{M}$ consists of all combinations of observer models and surprise along with the null model (NUL), namely $m \in \mathcal{M} = \{I_B(\text{BEL}), I_B(\text{BEL}_{\text{SI}}), I_B(\text{BEL}_{\text{SO}}), I_H(\text{BEL}), I_H(\text{BEL}_{\text{SI}}), I_H(\text{BEL}_{\text{SO}}), I_B(\text{PRE}), I_B(\text{PRE}_{\text{SI}}), I_B(\text{PRE}_{\text{SO}}), I_P(\text{PRE}), I_P(\text{PRE}_{\text{SI}}), I_P(\text{PRE}_{\text{SO}}), I_B(\text{OST}), I_B(\text{DIF}), I_B(\text{DIF}_{\text{OP}}), I_B(\text{DIF}_{\text{SP}}), I_P(\text{DIF}), I_P(\text{DIF}_{\text{OP}}), I_P(\text{DIF}_{\text{SP}}), \text{NUL}\}$, and the sequence $i_m(n)$ captures the surprise from model $m$. This section describes the composition of the model-independent data vector $\mathbf{y}_{\ell,c}$ and the model-specific design matrices $\mathbf{X}^{(1)}_{m,\ell,c}$ for each subject $\ell = \{1, \ldots, L = 16\}$ and experimental condition $c \in \mathcal{C} = \{\text{PcLc}, \text{PcLu}, \text{PuLc}, \text{PuLu}\}$ (see Sect. 4.2 for details on the experimental setup). The two-level GLM (see Sect. 2.3.1) is used with a single predictor per trial, simplifying (2.16) to

$$\mathbf{y}_{\ell,c} = \mathbf{x}^{(1)}_{m,\ell,c}\theta^{(1)}_{m,\ell,c} + \boldsymbol{\epsilon}^{(1)}_{m,\ell,c}$$

$$\theta^{(1)}_{m,\ell,c} = \epsilon^{(2)}_{m,\ell,c}. \tag{4.34}$$

The vector $\mathbf{y}_{\ell,c} = [\mathbf{y}^T_{\ell,c,b=1}, \ldots, \mathbf{y}^T_{\ell,c,b=50}]^T$ with $\mathbf{y}_{\ell,c,b} = [y''_{\ell,c,b}(n=1), \ldots, y''_{\ell,c,b}(n = N_{\ell,c,b})]^T$ contains the ERP amplitudes, which were normalized to zero mean and unit variance following

$$y''_{\ell,c,b}(n) = \frac{y_{\ell,c,b}(n) - \bar{y}_{\ell,c}}{\sigma_{y_{\ell,c}}}, \tag{4.35}$$

with subject- and condition-specific mean

$$\bar{y}_{\ell,c} = \frac{1}{\sum_{b=1}^{B} N_{\ell,c,b}} \sum_{b=1}^{B} \sum_{n=1}^{N_{\ell,c,b}} y_{\ell,c,b}(n) \tag{4.36}$$

and standard deviation

$$\sigma_{y_{\ell,c}} = \sqrt{\frac{1}{-1 + \sum_{b=1}^{B} N_{\ell,c,b}} \sum_{b=1}^{B} \sum_{n=1}^{N_{\ell,c,b}} (y_{\ell,c,b}(n) - \bar{y}_{\ell,c})^2}. \tag{4.37}$$

Note that while each subject perceived $n \in \{1, \ldots, 4\}$ stimuli on each block $b$, trials during which an artifact occurred or the wrong behavioral response was elicited are excluded from further analyses, leaving $n \in \{1, \ldots, N_{\ell,c,b}\}$ trials. Due to the very short block length, the normalization is done across blocks within one experimental condition and not within blocks as in Sect. 3.5. Analogous to (4.35), (4.36), and (4.37), the surprise values $i_{m,\ell,c,b}(n)$ are normalized to $i''_{m,\ell,c,b}(n)$ with zero mean and unit variance before constituting $\mathbf{x}^{(1)}_{m,\ell,c} = [\mathbf{i}^T_{m,\ell,c,b=1}, \ldots, \mathbf{i}^T_{m,\ell,c,b=50}]^T$ with $\mathbf{i}_{m,\ell,c,b} = [i''_{m,\ell,c,b}(n=1), \ldots, i''_{m,\ell,c,b}(n=N_{\ell,c,b})]^T$. After having $\mathbf{y}_{\ell,c}$ and $\mathbf{x}^{(1)}_{m,\ell,c}$ fully specified, model estimation commences as described in Sect. 2.4.2. Note that for the NUL model $\mathbf{i}_{\text{NUL},\ell,c,b} = [1, \ldots, 1]^T \in \mathbb{R}^{N_{\ell,c,b}}$ is an all-one vector as in Chap. 3 and Sect. 2.5.3.4.

The obtained experimental condition-specific variational free energies $F_{m,\ell,c}$ are summed to yield $F_{m,\ell} = \sum_{c \in C} F_{m,\ell,c}$. Based on fixed-effects assumptions (see Sect. 2.4.4), the group-level variational free energy is then given as $F_m = \sum_{\ell=1}^{L} F_{m,\ell}$. Finally, posterior model probabilities $P(m|\mathbf{y}) = \frac{e^{F_m}}{\sum_{\mu \in \mathcal{M}} e^{F_\mu}}$ are calculated.

The group log-Bayes factors $\log(\text{GBF})$ (2.58) (see Sects. 2.4.3 and 2.4.4) for the topographic maps and scalp maps (see Sect. 4.7.2 later on) were obtained by repeating the above model estimation for all electrodes $e \in \mathcal{E}$ and times $t \in [-100, 600]$ ms, with the amplitudes $y_{\ell,c,b,e}(n)$ being calculated from $y_{n,\ell,c,b,e}(t)$ after (1.5).

## 4.7  Results

This section first presents the conventional ERP analyses in the form of grand-average ERP waveforms for the ERP-specific virtual electrodes and the single electrodes which constitute the respective virtual electrodes. Furthermore, the signal-to-noise ratio estimates for all subjects and ERPs are shown. Next, the model-based trial-by-trial analyses present the posterior model probabilities for the whole model space. The model selection results are additionally displayed in the form of topographic maps and scalp maps, and the behavior of the $\text{BEL}_{\text{SI}}$ and $\text{PRE}_{\text{SI}}$ distributions is analyzed in more detail.

### 4.7.1  Conventional ERP Analyses

#### 4.7.1.1  Grand-Average ERP Waveforms

Figures 4.7, 4.8, 4.9, and 4.10 show grand-average ERP waveforms (4.2) separately for the more (($\bullet$), $k = 1$) and less (($\bullet$), $k = 2$) frequent ball colors in the certain prior (Pc, left panels) and uncertain prior (Pu, right panels) experimental conditions. The color of the waves matches the respective ball color, while the curve structure is defined by the experimental condition: (——) and (——) for certain likelihood, Lc, and

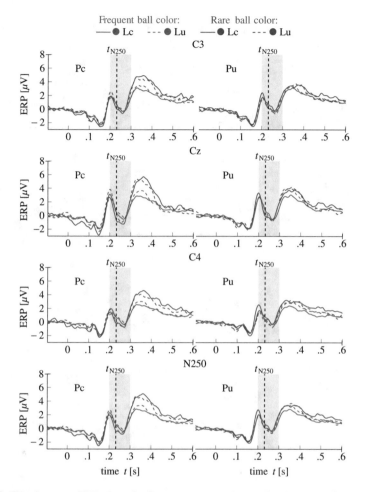

**Fig. 4.7** Grand-average ERP waves for frequent events (($\bullet$), (——) and (- - -)) and rare events (( $\bullet$), (——) and (- - -)) for certain likelihoods (Lc, (——) and (——)) and uncertain likelihoods (Lu, (- - -) and (- - -)) at electrodes C3, Cz, C4, and for the virtual electrode for the **N250**. The time interval for the search for maximum variance for the N250 is highlighted in *gray* and the point of time of maximum variance is marked by a *dashed black line* at all electrodes. *Left-hand panels* certain prior conditions (PcLc and PcLu). *Right-hand panels* uncertain prior conditions (PuLc and PuLu). The presence of a centrally distributed N250 wave in the latency range [200, 300] ms with $t_{N250} = 232$ ms is revealed

(- - -) and (- - -) for uncertain likelihood, Lu. The waves are shown for the N250, P3a, P3b, and SW, but in order to give a complete picture, the respective single electrodes C3, Cz, and C4 for the N250, Fz, FCz, and Cz for the P3a, and O1 and O2 for the SW, are displayed above the virtual electrodes. Note that for the P3b there is solely the electrode Pz. The maximum variance ERP analysis (see Sect. 4.2) reveals the presence of a centrally distributed N250 wave in the latency range [200, 300] ms

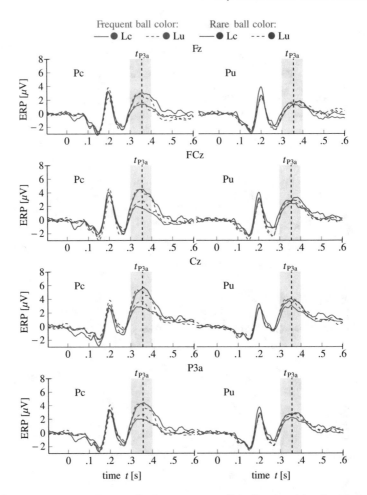

**Fig. 4.8** Grand-average ERP waves for frequent events ((●), ((——) and (- - -)) and rare events ((●), (——) and (- - -)) for certain likelihoods (Lc, (——) and (——)) and uncertain likelihoods (Lu, (- - -) and (- - -)) at electrodes Fz, FCz, Cz, and for the virtual electrode for the **P3a**. The time interval for the search for maximum variance for the P3a is highlighted in *gray* and the point of time of maximum variance is marked by a *dashed black line* at all electrodes. *Left-hand panels* certain prior conditions (PcLc and PcLu). *Right-hand panels* uncertain prior conditions (PuLc and PuLu). The presence of a frontally distributed P3a in the latency range [300, 400] ms with $t_{P3a} = 356$ ms is revealed

with $t_{N250} = 232$ ms (Fig. 4.7), a frontally distributed P3a (Fig. 4.8) in the latency range [300, 400] ms with $t_{P3a} = 356$ ms, a parietally distributed P3b (Fig. 4.9) in the latency range [300, 400] ms with $t_{P3b} = 380$ ms, and a posterior-positive SW in the latency range [400, 560] ms with $t_{SW} = 504$ ms (Fig. 4.10). The time intervals for the search for maximum variance are highlighted in gray and the time points of maximum variance are marked by a dashed black line at all electrodes.

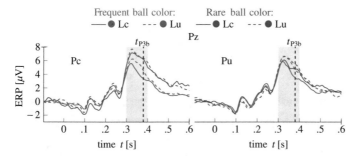

**Fig. 4.9** Grand-average ERP waves for frequent events ((●), (——) and (- - -)) and rare events ((●), (——) and (- - -)) for certain likelihoods (Lc, (——) and (——)) and uncertain likelihoods (Lu, (- - -) and (- - -)) at electrode Pz for the **P3b**. The time interval for the search for maximum variance for the P3b is highlighted in *gray* and the point of time of maximum variance is marked by a *dashed black line*. *Left-hand panels* certain prior conditions (PcLc and PcLu). *Right-hand panels* uncertain prior conditions (PuLc and PuLu). The presence of a parietally distributed P3b in the latency range [300, 400] ms with $t_{P3b} = 380$ ms is revealed

### 4.7.1.2  Signal-to-Noise Ratio Estimates

Tables 4.3, 4.4, 4.5, and 4.6 show subject-specific signal-to-noise ratio estimates $\widehat{SNR}$ [dB] (1.17) for the frequent (●) and rare (●) event types in all four experimental conditions $c \in C = \{PcLc, PcLu, PuLc, PuLu\}$. The average of the $\widehat{SNR}$ [dB] values over all experimental conditions and event types are presented in the rightmost column. The bottom row shows the average values for the respective columns highlighted in bold face. The total average $\widehat{SNR}$ [dB] over all subjects, event types, and experimental conditions is 0.69 dB for the N250, 0.56 dB for the P3a, 1.30 dB for the P3b, and 0.22 dB for the SW. These values show that the P3b is the strongest and most reliable of the four tested ERPs and that the SW is the most volatile, possibly causing the relative lack of formal understanding of its significance (Ruchkin et al. 1988; García-Larrea and Cézanne-Bert 1998; Spencer et al. 2001; Matsuda and Nittono 2015). Except for the SW, the $\widehat{SNR}_k$ [dB] are larger for the rare event type, which is in accordance with the results obtained in Sect. 3.6.1.2. The differing results for the SW are plausible when looking at the amplitudes of the grand-average ERP waves at $t_{SW}$ as displayed in Fig. 4.10. In contrast to all other ERPs (Figs. 4.7, 4.8, and 4.9), the amplitudes at $t_{SW}$ fluctuate around 0 $\mu$V and the SW is characterized by the spread of the ERP waves instead of their peak amplitudes.

The minimum $\widehat{SNR}$ [dB] values are slightly below 0 dB, with most values being between 0 dB and 1 dB. Taking the simulation results from Sect. 2.6 into account, these signal-to-noise ratios indicate that the data consist of *just* enough trials per subject to allow for reliable model selection results. As stated in Sect. 4.4.3, no training data set can be split off for the optimization of $\zeta$.

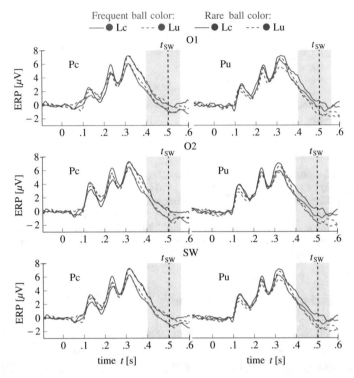

**Fig. 4.10** Grand-average ERP waves for frequent events ((●), (——) and (- - -)) and rare events ((●), (——) and (- - -)) for certain likelihoods (Lc, (——) and (——)) and uncertain likelihoods (Lu, (- - -) and (- - -)) at electrodes O1 and O2 for the **Slow Wave (SW)**. The time interval for the search for maximum variance for the SW is highlighted in *gray* and the point of time of maximum variance is marked by a *dashed black line* at all electrodes. *Left-hand panels* certain prior conditions (PcLc and PcLu). *Right-hand panels* uncertain prior conditions (PuLc and PuLu). The presence of a posterior-positive SW in the latency range [400, 560] ms with $t_{SW} = 504$ ms is revealed

### 4.7.2  Model-Based Trial-by-Trial Analyses

This section describes the model-based results. It starts with Table 4.7, which shows the posterior model probabilities of all tested probability distribution surprise combinations for the N250, P3a, P3b, and SW for the ERP-specific virtual electrodes and latencies, $t_{ERP}$, which were determined as described in Sect. 4.2. Next, Fig. 4.11 presents the group log-Bayes factors log (GBF) for all electrodes for the winning probability distribution surprise combinations over time for the late positive complex. It is followed by Fig. 4.12, which shows the respective scalp maps at the ERP-specific latencies. Figures 4.13, 4.14, 4.15 then detail the relations between the winning distribution-surprise combinations and the EEG data in the form of sequence-specific ERP waves. Last, Figs. 4.16 and 4.17 show the group log-Bayes factors over time and electrodes and as scalp map, respectively, for the superior model for the N250.

**Table 4.3** Subject-specific and average signal-to-noise ratio estimates $\widehat{\mathrm{SNR}}$ [dB] (1.17) for the **N250**

| Condition | PcLc | | PcLu | | PuLc | | PuLu | | All |
|---|---|---|---|---|---|---|---|---|---|
| Subject # | ● | ● | ● | ● | ● | ● | ● | ● | ● |
| 1 | 0.22 | 0.47 | 0.13 | −0.01 | 0.38 | 0.17 | −0.08 | −0.08 | 0.14 |
| 2 | 1.45 | 2.15 | 3.00 | 3.16 | 1.87 | 1.26 | 0.99 | 1.85 | 1.91 |
| 3 | 0.89 | 4.85 | 2.08 | 3.78 | 2.05 | 2.00 | 2.31 | 2.41 | 2.18 |
| 4 | 1.14 | 1.62 | 0.32 | 1.10 | 0.66 | 1.51 | 1.07 | 1.71 | 1.02 |
| 5 | 0.09 | −0.27 | −0.05 | 0.01 | 0.05 | 0.15 | 0.01 | 0.06 | 0.03 |
| 6 | 0.17 | 1.33 | 0.37 | 0.08 | 1.05 | 1.20 | 0.79 | 1.18 | 0.66 |
| 7 | 0.12 | −0.23 | −0.06 | 0.23 | 0.29 | 0.07 | −0.05 | 0.09 | 0.07 |
| 8 | 0.00 | 0.15 | −0.06 | −0.14 | −0.06 | −0.07 | −0.10 | 0.10 | −0.02 |
| 9 | 1.92 | 1.18 | 2.46 | 1.84 | 1.82 | 1.15 | 2.16 | 2.17 | 1.94 |
| 10 | 1.45 | 0.37 | 2.09 | 2.05 | 0.87 | 0.16 | 0.26 | 0.44 | 1.04 |
| 11 | 0.17 | −0.27 | 0.03 | −0.10 | 0.69 | −0.12 | −0.06 | 0.15 | 0.14 |
| 12 | 1.02 | 0.38 | 0.70 | 0.33 | −0.08 | −0.04 | −0.06 | −0.08 | 0.34 |
| 13 | −0.03 | 0.28 | −0.01 | 0.13 | 0.25 | 0.50 | −0.07 | 0.02 | 0.09 |
| 14 | 0.47 | −0.16 | 0.35 | 0.21 | 0.10 | 0.87 | 0.89 | 0.86 | 0.47 |
| 15 | 0.44 | 0.82 | 0.27 | 1.54 | 0.66 | 2.51 | 0.71 | 2.61 | 1.04 |
| 16 | −0.04 | −0.13 | −0.05 | −0.12 | −0.07 | −0.03 | 0.21 | −0.09 | −0.02 |
| All | **0.59** | **0.78** | **0.72** | **0.88** | **0.66** | **0.70** | **0.56** | **0.84** | **0.69** |

The values are shown separately for the frequent (●) and rare (●) event types in all four experimental conditions $c \in \mathcal{C} = \{$PcLc, PcLu, PuLc, PuLu$\}$. The average values over both types of events (●) and all experimental conditions are presented in the rightmost column. The bottom row shows the averaged estimated signal-to-noise ratios for the respective columns

#### 4.7.2.1 Posterior Model Probabilities

Table 4.7 displays the posterior model probabilities $P(m|\mathbf{y})$ (2.59) for all tested distribution surprise combinations separately for N250 (electrodes C3, Cz, C4), P3a (electrodes Fz, FCz, Cz), P3b (electrode Pz), and SW (electrodes O1, O2) latencies. Trial-by-trial amplitude variability for the P3a is clearly best accounted for by the $\mathrm{BEL_{SI}}$ distribution with Bayesian surprise ($P(m = I_B(\mathrm{BEL_{SI}})|\mathbf{y}) > 0.99$). The $\mathrm{PRE_{SI}}$ distribution with predictive surprise shows maximum posterior model probability for P3b amplitudes ($P(I_P(\mathrm{PRE_{SI}})|\mathbf{y}) > 0.99$), while for SW amplitudes the $\mathrm{PRE_{SI}}$ distribution with postdictive surprise has the highest posterior model probability ($P(I_B(\mathrm{PRE_{SI}})|\mathbf{y}) = 0.95$). Predictive surprise based on the DIF model shows the highest posterior model probability ($P(I_P(\mathrm{DIF})|\mathbf{y}) > 0.99$) for the N250. All posterior model probabilities for entropy calculated over any of the belief

**Table 4.4** Subject-specific and average signal-to-noise ratio estimates $\widehat{\text{SNR}}$ [dB] (1.17) for the **P3a**

| Condition | PcLc | | PcLu | | PuLc | | PuLu | | All |
|---|---|---|---|---|---|---|---|---|---|
| Subject # | ● | ● | ● | ● | ● | ● | ● | ● | ● |
| 1 | −0.02 | 1.47 | −0.01 | 0.08 | 0.14 | −0.12 | 0.18 | −0.06 | 0.08 |
| 2 | 0.22 | 1.38 | 0.14 | −0.13 | 0.29 | 0.12 | 0.48 | 0.68 | 0.31 |
| 3 | 0.10 | −0.14 | −0.05 | −0.13 | 0.22 | 0.07 | −0.01 | −0.08 | 0.02 |
| 4 | −0.02 | 0.25 | 0.46 | 0.88 | 0.26 | 1.14 | 0.95 | 0.56 | 0.51 |
| 5 | −0.05 | 0.50 | −0.07 | 0.03 | −0.06 | −0.07 | 0.47 | −0.06 | 0.05 |
| 6 | 0.79 | 1.29 | 0.90 | 1.74 | 0.18 | 0.19 | 0.12 | 0.46 | 0.60 |
| 7 | 3.50 | 1.40 | 3.06 | 3.37 | 3.83 | 4.16 | 4.33 | 2.74 | 3.48 |
| 8 | 0.54 | 1.02 | 1.33 | 3.39 | 0.64 | 2.02 | 0.30 | 0.96 | 1.09 |
| 9 | 1.18 | 1.20 | 0.22 | −0.13 | 0.37 | 0.33 | 0.15 | −0.05 | 0.43 |
| 10 | −0.05 | 0.98 | 0.04 | 0.75 | −0.07 | −0.05 | 0.17 | −0.07 | 0.14 |
| 11 | −0.01 | −0.21 | 0.03 | −0.13 | 0.12 | −0.11 | −0.06 | −0.10 | −0.02 |
| 12 | 0.01 | −0.03 | 0.03 | −0.06 | 1.00 | 0.77 | 0.37 | 0.41 | 0.32 |
| 13 | 0.10 | 0.95 | 0.57 | 0.93 | 0.06 | −0.11 | 0.93 | 0.47 | 0.43 |
| 14 | 0.03 | 0.36 | 1.48 | 1.47 | 0.20 | 0.82 | 1.09 | 2.12 | 0.84 |
| 15 | −0.05 | −0.03 | −0.08 | −0.04 | 0.33 | 0.11 | −0.03 | −0.12 | 0.02 |
| 16 | 0.24 | 1.12 | 0.42 | 1.98 | 0.13 | 0.09 | 0.98 | 2.02 | 0.71 |
| All | **0.41** | **0.72** | **0.53** | **0.88** | **0.48** | **0.59** | **0.65** | **0.62** | **0.56** |

The values are shown separately for the frequent (●) and rare (●) event types in all four experimental conditions $c \in C = \{$PcLc, PcLu, PuLc, PuLu$\}$. The average values over both types of events (●) and all experimental conditions are presented in the rightmost column. The bottom row shows the averaged estimated signal-to-noise ratios for the respective columns

distributions remain negligible. These results imply that the $\text{BEL}_{\text{SI}}$ distribution with Bayesian surprise provides superior predictions with regard to trial-by-trial P3a variability at fronto-central electrodes. In addition to that, the $\text{PRE}_{\text{SI}}$ distribution with predictive surprise is superior with regard to centro-parietal P3b amplitude variability, and the $\text{PRE}_{\text{SI}}$ distribution with postdictive surprise is superior with regard to occipito-parietal SW amplitude variability.

### 4.7.2.2　Detailed Results for the Late Positive Complex

Figure 4.11 displays group log-Bayes factors log $(\text{GBF}_{m \to \text{NUL}})$ with $m \in \{I_B(\text{BEL}_{\text{SI}}),$ $I_B(\text{BEL}), I_B(\text{PRE}_{\text{SI}}), I_P(\text{PRE}_{\text{SI}})\}$ of the winning $\text{BEL}_{\text{SI}}$ and $\text{PRE}_{\text{SI}}$ distributions and of the BEL distribution versus the NUL model for the whole epoch length ($[-100, 600]$ ms around stimulus presentation) and all electrodes. The highest group

**Table 4.5**  Subject-specific and average signal-to-noise ratio estimates $\widehat{\mathrm{SNR}}$ [dB] (1.17) for the **P3b**

| Condition | Signal-to-noise ratio estimate | | | | | | | | |
|---|---|---|---|---|---|---|---|---|---|
|  | PcLc | | PcLu | | PuLc | | PuLu | | All |
| Subject # | ● | ● | ● | ● | ● | ● | ● | ● | ● |
| 1 | 1.10 | 3.17 | 1.62 | 2.18 | 0.42 | 0.93 | 0.52 | 0.86 | 1.12 |
| 2 | 0.26 | 0.96 | 0.07 | 1.13 | 0.47 | 0.40 | 0.75 | 1.78 | 0.61 |
| 3 | −0.06 | 0.38 | 0.80 | 1.88 | 0.30 | 0.28 | 0.48 | −0.07 | 0.43 |
| 4 | −0.05 | 0.22 | 0.94 | 1.25 | 0.16 | 1.20 | 0.33 | 0.60 | 0.50 |
| 5 | 0.15 | 0.62 | 0.19 | 0.73 | −0.03 | −0.12 | 0.55 | 1.04 | 0.32 |
| 6 | 1.82 | 5.30 | 2.11 | 5.10 | 2.63 | 2.72 | 2.18 | 1.75 | 2.49 |
| 7 | 3.75 | 2.19 | 3.91 | 4.91 | 4.83 | 5.28 | 4.73 | 3.86 | 4.28 |
| 8 | 0.56 | 0.13 | 1.93 | 2.55 | 1.27 | 2.12 | 1.46 | 0.20 | 1.18 |
| 9 | 1.81 | 1.99 | 0.79 | 0.39 | 1.21 | 1.56 | 1.20 | 0.28 | 1.16 |
| 10 | 0.04 | 1.00 | 0.17 | 0.31 | 0.06 | −0.11 | −0.07 | −0.08 | 0.11 |
| 11 | 0.83 | 3.60 | 1.22 | 2.19 | 2.26 | 3.14 | 2.02 | 3.13 | 1.97 |
| 12 | 0.69 | 0.50 | 0.88 | 2.71 | 1.82 | 0.86 | 1.61 | 2.26 | 1.37 |
| 13 | 0.71 | 3.39 | 1.22 | 1.09 | 1.56 | 0.92 | 1.92 | 2.20 | 1.45 |
| 14 | 0.22 | 1.04 | 2.32 | 2.49 | 0.70 | 0.80 | 1.20 | 2.87 | 1.31 |
| 15 | 0.60 | 1.40 | 0.26 | 1.86 | 0.42 | 2.44 | 0.64 | 1.08 | 0.92 |
| 16 | 0.65 | 1.95 | 1.39 | 3.49 | 0.99 | 1.71 | 1.58 | 2.71 | 1.57 |
| All | **0.82** | **1.74** | **1.24** | **2.14** | **1.19** | **1.51** | **1.32** | **1.53** | **1.30** |

The values are shown separately for the frequent (●) and rare (●) event types in all four experimental conditions $c \in C = \{\mathrm{PcLc}, \mathrm{PcLu}, \mathrm{PuLc}, \mathrm{PuLu}\}$. The average values over both types of events (●) and all experimental conditions are presented in the rightmost column. The bottom row shows the averaged estimated signal-to-noise ratios for the respective columns

log-Bayes factors (red traces) represent better fits between surprise and the ERP amplitudes. Bayesian updating and predictive surprise seem to provide accurate approximations to the actual data, with a fronto-central focus within the P3a latency range, a centro-parietal focus in the P3b latency range, and an occipito-parietal focus in the SW latency range. The group log-Bayes factors for the BEL distribution are shown to illustrate how specific the winning models fit the topography of the late positive complex. Note how for the BEL distribution the topography is virtually flat.

Figure 4.12 displays group log-Bayes factors in the form of scalp maps separately for the P3a, P3b, and SW. Note that these maps do *not* show scalp distributions of measured data, but the degree to which different kinds of surprise, which are calculated based on the distributions, approximate measured trial-by-trial ERP amplitudes. The P3a maps show a circumscribed fronto-central focus along with a left-occipital spot. For the P3b, a more posteriorly (centro-parietal) and broadly distributed focus appears. Finally, the fit between surprise and measured data is sharply confined to the occipito-parietal region with regard to the SW. While the maps for the different

**Table 4.6** Subject-specific and average signal-to-noise ratio estimates $\widehat{\text{SNR}}$ [dB] (1.17) for the **SW**

| Condition | Signal-to-noise ratio estimate | | | | | | | | |
|---|---|---|---|---|---|---|---|---|---|
| | PcLc | | PcLu | | PuLc | | PuLu | | All |
| Subject # | ● | ● | ● | ● | ● | ● | ● | ● | ● |
| 1 | 1.12 | −0.32 | 0.39 | −0.08 | 0.21 | 0.21 | −0.07 | −0.10 | 0.32 |
| 2 | 0.45 | −0.23 | 0.56 | −0.06 | 0.80 | 0.12 | 0.07 | 0.21 | 0.33 |
| 3 | 2.31 | 0.15 | 0.72 | 0.15 | 0.44 | −0.08 | 0.91 | 0.98 | 0.90 |
| 4 | −0.05 | −0.17 | −0.07 | 0.07 | −0.05 | −0.05 | −0.05 | 0.08 | −0.03 |
| 5 | −0.04 | −0.14 | 0.45 | 0.34 | 0.11 | −0.14 | 0.19 | 0.79 | 0.20 |
| 6 | −0.05 | −0.14 | −0.06 | −0.16 | 0.25 | −0.05 | −0.07 | 0.28 | 0.02 |
| 7 | 0.34 | −0.23 | −0.01 | 1.01 | −0.03 | −0.06 | −0.07 | 0.28 | 0.14 |
| 8 | −0.04 | −0.09 | 0.27 | 0.86 | −0.04 | −0.16 | 1.68 | 2.49 | 0.64 |
| 9 | 0.03 | 1.15 | 0.36 | −0.07 | −0.07 | 0.04 | −0.06 | −0.10 | 0.08 |
| 10 | −0.06 | −0.11 | 0.06 | 0.29 | −0.08 | −0.10 | 0.16 | 0.11 | 0.03 |
| 11 | −0.04 | 0.22 | 0.05 | 0.17 | −0.05 | 0.13 | 0.42 | 0.24 | 0.11 |
| 12 | 0.36 | −0.08 | −0.07 | 0.02 | 0.12 | −0.09 | −0.06 | −0.11 | 0.05 |
| 13 | 0.04 | 0.22 | 0.30 | −0.15 | −0.06 | 0.06 | 0.02 | −0.13 | 0.05 |
| 14 | 0.47 | −0.30 | 0.62 | 0.43 | 0.03 | −0.01 | −0.01 | 0.31 | 0.26 |
| 15 | 0.74 | −0.19 | 0.11 | 0.03 | 0.02 | −0.04 | 0.04 | −0.1 | 0.14 |
| 16 | 0.25 | −0.02 | 0.35 | 0.66 | 0.11 | 0.20 | −0.02 | 0.22 | 0.21 |
| All | **0.36** | **-0.02** | **0.25** | **0.22** | **0.11** | **0.00** | **0.19** | **0.34** | **0.22** |

The values are shown separately for the frequent (●) and rare (●) event types in all four experimental conditions $c \in C = \{\text{PcLc, PcLu, PuLc, PuLu}\}$. The average values over both types of events (●) and all experimental conditions are presented in the rightmost column. The bottom row shows the averaged estimated signal-to-noise ratios for the respective columns

models resemble each other for each ERP, this is due to limits of the graphical representation and similar behavior of the models, which will be shown later on in this chapter.

### 4.7.2.3  Trial-by-Trial Behavior of the $\text{BEL}_{\text{SI}}$ and $\text{PRE}_{\text{SI}}$ Distributions

**Detailed Analysis of the $\text{BEL}_{\text{SI}}$ Distribution**

Figure 4.13 shows the correlation between Bayesian surprise, $I_B(n) = D_{\text{KL}}(P_{\mathcal{U}}(n-1) \,||\, P_{\mathcal{U}}(n))$ (1.19), that was obtained from the $\text{BEL}_{\text{SI}}$ distribution (4.24) (with hyperparameter $\zeta = 0.65$) and neural activity (here summarized through sequence-specific ERP waves at electrode FCz) across the four experimental conditions $c \in \{\text{PcLc} (\text{—+—}), \text{PcLu} (\text{- ▲ -}), \text{PuLc} (\text{—▼—}), \text{PuLu} (\text{- ■ -})\}$, and across eight potential sequences of three successive ball stimuli (cf. (●●●) upper right panels—(●●●) lower right panels). Note that while Bayesian surprise is shown at each stage of the sequences on

**Table 4.7** Posterior model probabilities for all tested distribution-surprise combinations for the N250 ($t_{N250} = 232$ ms), for the P3a ($t_{P3a} = 356$ ms), for the P3b ($t_{P3b} = 370$ ms), and for the SW ($t_{SW} = 504$ ms)

| Surprise | Distribution | ERP | | | |
|---|---|---|---|---|---|
| | | N250 | P3a | P3b | SW |
| $I_B$ | BEL | < 0.01 | < 0.01 | < 0.01 | < 0.01 |
| $I_B$ | BEL$_{SI}$ | < 0.01 | > **0.99** | < 0.01 | < 0.01 |
| $I_B$ | BEL$_{SO}$ | < 0.01 | < 0.01 | < 0.01 | < 0.01 |
| $I_H$ | BEL | < 0.01 | < 0.01 | < 0.01 | < 0.01 |
| $I_H$ | BEL$_{SI}$ | < 0.01 | < 0.01 | < 0.01 | < 0.01 |
| $I_H$ | BEL$_{SO}$ | < 0.01 | < 0.01 | < 0.01 | < 0.01 |
| $I_P$ | PRE | < 0.01 | < 0.01 | < 0.01 | < 0.01 |
| $I_P$ | PRE$_{SI}$ | < 0.01 | < 0.01 | > **0.99** | 0.05 |
| $I_P$ | PRE$_{SO}$ | < 0.01 | < 0.01 | < 0.01 | < 0.01 |
| $I_B$ | PRE | < 0.01 | < 0.01 | < 0.01 | < 0.01 |
| $I_B$ | PRE$_{SI}$ | < 0.01 | < 0.01 | < 0.01 | **0.95** |
| $I_B$ | PRE$_{SO}$ | < 0.01 | < 0.01 | < 0.01 | < 0.01 |
| $I_B$ | DIF | < 0.01 | < 0.01 | < 0.01 | < 0.01 |
| $I_B$ | DIF$_{OP}$ | < 0.01 | < 0.01 | < 0.01 | < 0.01 |
| $I_B$ | DIF$_{SP}$ | < 0.01 | < 0.01 | < 0.01 | < 0.01 |
| $I_P$ | DIF | > **0.99** | < 0.01 | < 0.01 | < 0.01 |
| $I_P$ | DIF$_{OP}$ | < 0.01 | < 0.01 | < 0.01 | < 0.01 |
| $I_P$ | DIF$_{SP}$ | < 0.01 | < 0.01 | < 0.01 | < 0.01 |
| | NUL | < 0.01 | < 0.01 | < 0.01 | < 0.01 |

the left, the ERP waves, shown on the right, are in response to the third ball stimulus only. The electrode FCz was chosen as it represents the center of the region of interest for the P3a. A close correlation between Bayesian surprise and neural activities is revealed by comparing Bayesian surprise for the third trial, $I_B(n = 3)$, and the corresponding ERP waves in the typical P3a latency range of 300 to 400 ms which is highlighted in gray. Specifically, gradually increasing ERP wave amplitudes are associated with successive increases in Bayesian surprise, $I_B(n = 3)$ (compare (●●●) vs. (●●●) vs. (●●●) vs. (●●●) and (●●●) vs. (●●●) vs. (●●●) vs. (●●●), respectively). Further, the left panels show that Bayesian surprise is mainly grouped by likelihood (Lc (——) vs. Lu (---)). In order to show this effect in the ERP data, the waves for certain (Lc) and uncertain (Lu) likelihood have been averaged separately, regardless of prior probabilities. The ERP waves also seem to reflect the degree to which Bayesian surprise $I_B(n = 3)$ under Lc conditions (——) surpasses $I_B(n = 3)$ under Lu conditions (---).

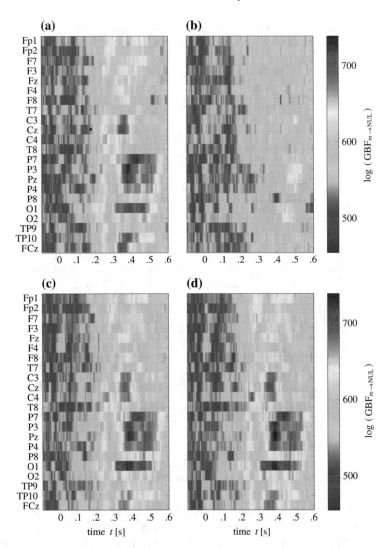

**Fig. 4.11** Topographic maps which show the degree to which the models $m \in \{I_B(\mathrm{BEL_{SI}}), I_B(\mathrm{BEL}), I_B(\mathrm{PRE_{SI}}), I_P(\mathrm{PRE_{SI}})\}$ approximate measured trial-by-trial ERP amplitudes in group log-Bayes factors $\log(\mathrm{GBF}_{m \to \mathrm{NUL}})$ over electrodes and time. **a** $\log(\mathrm{GBF}_{I_B(\mathrm{BEL_{SI}}) \to \mathrm{NUL}})$. **b** $\log(\mathrm{GBF}_{I_B(\mathrm{BEL}) \to \mathrm{NUL}})$. **c** $\log(\mathrm{GBF}_{I_B(\mathrm{PRE_{SI}}) \to \mathrm{NUL}})$. **d** $\log(\mathrm{GBF}_{I_P(\mathrm{PRE_{SI}}) \to \mathrm{NUL}})$

## Detailed Analysis of the PRE$_{\mathrm{SI}}$ Distribution

Figures 4.14 and 4.15 show the correlation between predictive surprise, $I_P(n)$ (1.20), and postdictive surprise, $I_B(n)$ (1.18), respectively, that was obtained from the PRE$_{\mathrm{SI}}$ distribution (4.25) (with hyper-parameter $\zeta = 0.65$) and neural activities in form of

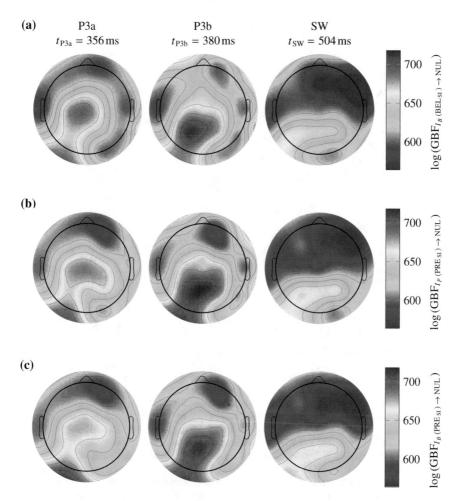

**Fig. 4.12** Scalp maps of group log-Bayes factors at $t_{P3a} = 356$ ms (*left column*), $t_{P3b} = 380$ ms (*middle column*), and $t_{SW} = 504$ ms (*right column*) for the **P3a**, **P3b**, and **SW**, respectively. Notice central foci within P3a latency and occipito-parietal foci in the P3b and SW latency. The P3a maps show a circumscribed fronto-central focus along with a left-occipital spot. For the P3b, a more posteriorly (occipital-parietal) and broadly distributed focus appears. Finally, the fit between surprise and measured data is sharply confined to the left-occipital region with regard to the SW. **a** $\log\left(\text{GBF}_{I_B(\text{BEL}_{SI})\to\text{NUL}}\right)$. **b** $\log\left(\text{GBF}_{I_P(\text{PRE}_{SI})\to\text{NUL}}\right)$. **c** $\log\left(\text{GBF}_{I_B(\text{PRE}_{SI})\to\text{NUL}}\right)$

sequence-specific ERP waves across the four experimental conditions $c \in \{$PcLc $(\text{--}\!\!+\!\!\text{--})$, PcLu $(\text{-}\!\!\blacktriangle\!\!\text{-})$, PuLc $(\text{--}\!\!\blacktriangledown\!\!\text{--})$, PuLu $(\text{-}\!\!\blacksquare\!\!\text{-})\}$, and across eight potential sequences of three successive ball stimuli (cf. ($\bullet\bullet\bullet$) upper right panels—($\bullet\bullet\bullet$) lower right panels). Figure 4.14 shows the ERP waves obtained at electrode Pz, as this is the single electrode of interest for the P3b, while Fig. 4.15 shows the waves for the electrode O1, as Fig. 4.12c indicates strongest postdictive surprise-related activation

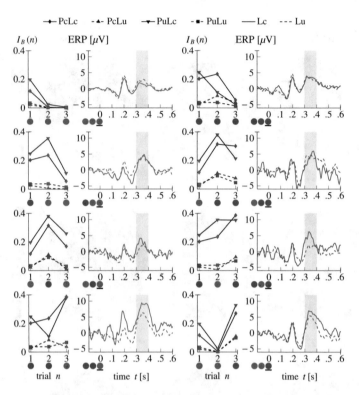

**Fig. 4.13** Relationships between Bayesian surprise based on the **BEL$_{SI}$** distribution with hyper-parameter $\zeta = 0.65$ and the ERP amplitudes across sequences of observed events. *Left panels* Bayesian surprise $I_B(n)$ (1.19) over trials $n = 1, 2, 3$ for all probability conditions PcLc (-◆-), PcLu (-▲-), PuLc (-▼-), and PuLu (-■-). The sequence of observed events is shown below each figure with (●) denoting a frequent and (●) a rare event. A clear likelihood effect is visible ((——) vs. (- - -)). *Right panels* In order to show this effect in the ERP amplitudes the waves for certain (Lc, ( —— )) and uncertain (Lu, (- - -)) likelihood have been averaged separately, yet regardless of prior probabilities. These ERP waves are shown for the third observation of a sequence, $o(n=3)$, at electrode **FCz**. Gradually increasing ERP wave amplitudes are mirrored by successive increases in Bayesian surprise $I_B(n=3)$

at this electrode for the SW. Note that while surprise is shown at each stage of the sequences on the left, the ERP waves, shown on the right, are in response to the third ball stimulus only.

As for Bayesian surprise based on the BEL$_{SI}$ distribution in Fig. 4.13, a close correlation between predictive surprise and postdictive surprise based on the PRE$_{SI}$ distribution and neural activities is revealed by comparing the surprise values for the third trial with the corresponding ERP waves in the ERP-typical latency ranges, which are highlighted in gray. Specifically, gradually increasing ERP wave amplitudes are associated with successive increases in surprise (compare (●●●) vs. (●●●) vs. (●●●) vs. (●●●) and (●●●) vs. (●●●) vs. (●●●) vs. (●●●), respectively). Further-

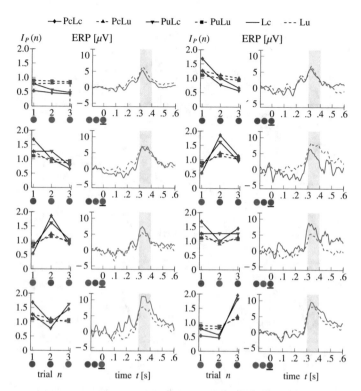

**Fig. 4.14** Relationships between predictive surprise based on the **PRE$_{SI}$** distribution with hyperparameter $\zeta = 0.65$ and the ERP amplitudes across sequences of observed events. *Left panels* Predictive surprise $I_P(n)$ (1.20) over trials $n = 1, 2, 3$ for all probability conditions PcLc (—+—), PcLu (- ▲ -), PuLc (—▼—), and PuLu (- ■ -). The sequence of observed events is shown below each figure with (●) denoting a frequent and (●) a rare event. A clear likelihood effect is visible ((——) vs. (- - -)). *Right panels* In order to show this effect in the ERP amplitudes the waves for certain (Lc, (——)) and uncertain (Lu, (- - -)) likelihood have been averaged separately, yet regardless of prior probabilities. These ERP waves are shown for the third observation of a sequence $o(n=3)$ at electrode **Pz**. Gradually increasing ERP wave amplitudes are mirrored by successive increases in predictive surprise $I_P(n=3)$

more, the left panels show that surprise for the PRE$_{SI}$ distribution is mainly grouped by likelihood (Lc (——) vs. Lu (- - -)), same as Bayesian surprise based on the BEL$_{SI}$ distribution shown in Fig. 4.13. In order to show this effect in the ERP amplitudes, the waves for certain (Lc) and uncertain (Lu) likelihood have been averaged separately, regardless of prior probabilities. These ERP waves reflect the degree to which surprise under Lc conditions (——) surpasses surprise under Lu conditions (- - -). Specifically, the differences between surprise under Lc and Lu conditions at $n = 3$ and the two ERP waves in the highlighted latency ranges seem to mirror each other.

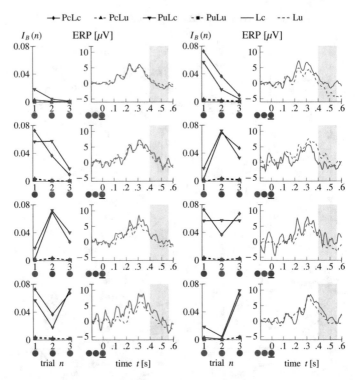

**Fig. 4.15** Relationships between postdictive surprise based on the **PRE$_{SI}$** distribution with hyperparameter $\zeta = 0.65$ and the ERP amplitudes across sequences of observed events. *Left panels* Postdictive surprise $I_B(n)$ (1.18) over trials $n = 1, 2, 3$ for all probability conditions PcLc (-+-), PcLu (-▲-), PuLc (-▼-), and PuLu (-■-). The sequence of observed events is shown below each figure with (●) denoting a frequent and (●) a rare event. A clear likelihood effect is visible ((——) vs. (- - -)). *Right panels* In order to show this effect in the ERP amplitudes the waves for certain (Lc, (——)) and uncertain (Lu, (- - -)) likelihood have been averaged separately, yet regardless of prior probabilities. These ERP waves are shown for the third observation of a sequence $o(n = 3)$ at electrode **O1**. Gradually increasing ERP wave amplitudes are mirrored by successive increases in predictive surprise $I_P(n = 3)$

### 4.7.2.4   Detailed Results for the N250

The N2-P3 complex has long been considered as an index of the operation of adaptive brain systems that enable the brain to anticipate the occurrence of sensory input and to react to unexpected events (Hillyard and Picton 1987). Several varieties of the fronto-centrally distributed N2 have been reported in (Folstein and Van Petten 2008). Towey et al. (1980) showed an increase in N250 latency with increased difficulty in discriminating target from non-target auditory stimuli, and they consequently associated N250 latency with decision latency. This knowledge warrants attention for the N250 when examining probabilistic inference.

Figure 4.16 shows the degree to which surprise as calculated by the DIF model approximates measured trial-by-trial ERP amplitudes in group log-Bayes factors. Figure 4.17 shows group log-Bayes factors in the form of scalp maps for the N250 latency. For the N250, a central, slightly right-lateralized spot becomes apparent. Table 4.7 shows that ERP amplitude fluctuations of the fronto-centrally distributed N250 (Towey et al. 1980) could be best accounted for by predictive surprise based on the DIF model with a uniform initial prior with the maximum posterior model probability $P(I_P(\text{DIF})|\mathbf{y}) > 0.99$.

The DIF model basically rests on counting observed events with short-term and long-term exponential forgetting rates (see Chap. 3). Thus, this model envisages probabilistic inference on the urn-ball task to be based upon memory for the frequency of occurrence of observable events without any prior knowledge. A detailed analysis of the DIF model is omitted at this point as it has been thoroughly discussed in Chap. 3.

**Fig. 4.16** Topographic map which shows the degree to which predictive surprise based on the **DIF** model approximates measured trial-by-trial ERP amplitudes in group log-Bayes factors $\log\left(\text{GBF}_{I_P(\text{DIF})\rightarrow\text{NUL}}\right)$ over electrodes and time

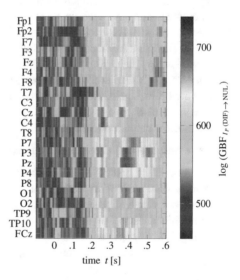

**Fig. 4.17** Scalp map of group log-Bayes factors $\log\left(\text{GBF}_{I_P(\text{DIF})\rightarrow\text{NUL}}\right)$ at time $t_{\text{N250}} = 232\,\text{ms}$ for the **N250**. A central focus becomes apparent along with a left-occipital spot

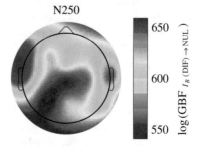

## 4.8   Summary and Discussion

This chapter explored neural correlates of Bayesian inference by combining the urn-ball task (Fig. 4.2) with computational modeling of trial-by-trial ERP amplitudes in the form of a Bayesian observer model. This approach led to the discovery that dissociable cortical signals seem to code and compute distinguishable aspects of Bayes-optimal probabilistic inference. Thus, the ERP components composing the late positive complex could be dissociated with regard to their presumed function in Bayesian inference (cf. Figs. 4.11 and 4.12, Table 4.7). Specifically, the late positive complex was decomposed spatially, temporally and functionally into three separable ERP components (see also Dien et al. 2004): (1) Bayesian surprise yielded superior approximations of activation changes in anteriorly distributed P3a waves at relatively short latency. (2) Postdictive surprise best explains posteriorly distributed SW amplitudes at latest latency. (3) Predictive surprise outperformed Bayesian surprise with regard to activation changes in parietally distributed P3b waves at intermediate latency. As a whole, these results are consistent with the Bayesian brain hypothesis insofar as dissociable neural activities seem to code and compute various aspects of Bayesian inference modulated via the hyper-parameter $\zeta$. The N250 was best explained by the digital filtering model, suggesting a dissociation between fast, memory-based (N250) and slow, model-based (late positive complex) forms of surprise in the brain. The distinction between memory-based and model-based surprise may be viewed in the context of more general distinctions, such as high-level and low-level error detection (Strauss et al. 2015).

**Surprise and the Late Positive Complex**

Bayesian updating generally reflects the Kullback–Leibler divergence between two probability distributions as defined in (1.18) and (1.19). The difference between Bayesian surprise and postdictive surprise should be shortly recapitulated for the functional dissociation of the late positive complex: Bayesian surprise represents the change in beliefs over hidden states given new observations, which equals the Kullback–Leibler divergence between $P_{\mathcal{U}}(n-1)$ and $P_{\mathcal{U}}(n)$ (see (4.20)). In contrast, postdictive surprise represents the change in predictions over future events given new observations and equals the Kullback–Leibler divergence between $P_{\mathcal{K}}(n)$ and $P_{\mathcal{K}}(n+1)$ (see (4.22)). Predictive surprise is simply the surprise over the current observation under its degree of prediction (see (4.23)). For a more detailed overview on surprise see Sect. 1.4. The here obtained results imply that Bayesian surprise is related to trial-by-trial P3a amplitude variability, postdictive surprise suitably models trial-by-trial SW amplitude variability, and predictive surprise best predicts trial-by-trial P3b amplitude variability.

**Detailed Discussion of the Bayesian Observer**

The urn-ball task was specifically designed to examine the neural bases of Bayesian inference. The electrophysiological findings suggest that the brain acts as a Bayesian observer, i.e., that it might adjust an internal model of the environment, which consists of beliefs about hidden states and predictions of observable events in the environment.

This internal model enables inference, and the beliefs provide an abstract explanation of adaptive cognition and behavior, which has been instantiated in schemes like the free-energy principle (Friston 2010; Lieder et al. 2013). Furthermore, the predictions of future events provide an abstract explanation of preadaptive cognition and behavior (Fuster 2014).

Against the background that the P3a originates from prefrontal cortical regions whilst the P3b is generated in temporal/parietal regions (Polich 2007), these results suggest how a network of brain areas may enable Bayesian inference. Specifically, while belief updating and prediction updating seem to be computed in prefrontal cortical regions (Lee et al. 2007), predictive surprise seems to originate from parietal regions (de Lange et al. 2010; d'Acremont et al. 2013; Kira et al. 2015). The occipital scalp topography of the SW requires a short comment. One plausible possibility is that the SW reflects the setting and updating of pre-adaptive biases. Kok and colleagues found that perceptual predictions trigger the formation of specific stimulus templates in the primary visual cortex to efficiently process sensory inputs (Kok et al. 2014).

**The Impact of Probability Weighting**

The hyper-parameter $\zeta$ was fitted by minimizing the mean squared error to approximate optimal decision behavior. This approach provided $\zeta = 0.65$, with $\zeta < 1$ being associated with inverse S-shaped probability weighting (Fig. 1.5). A Bayesian observer with weighting of the inference input was compared with an otherwise equivalent observer without probability weighting. The observer with probability weighting outperformed the unweighting observer when predicting observed ERPs (Table 4.7). These findings seem to demonstrate a ubiquitous role of probability weighting in probabilistic inference (Kahneman and Tversky 1979; Tversky and Kahneman 1992; Fox and Poldrack 2009). The alternative possibility that the nonlinearity might lie at the level of mapping from probability distributions to electrophysiological responses such that the electrophysiological responses may be a nonlinear function of the neuronal representation of unweighted probabilities was not supported by the data.

The primary effect of inverse S-shaped probability weighting (with $\zeta < 1$) on Bayesian updating is increased *uncertainty*. An inspection of Fig. 4.5 reveals that all posterior probabilities based on an observer with probability weighting lie between the corresponding posterior probabilities based on an observer without probability weighting and $P(q = 1|o = k) = 0.5$, which equals the point of maximum uncertainty. Probability weighting might constitute one of the reasons why empirical support for the Bayesian brain hypothesis (Knill and Pouget 2004; Friston 2005; Doya et al. 2007; Gold and Shadlen 2007; Kopp 2008) has apparently been so difficult to obtain in former studies. Note that earlier attempts to identify brain areas that weight probabilities did not lead to converging results; yet, common denominators of potential areas seem to lie within fronto-striatal loops (Trepel et al. 2005; Preuschoff et al. 2006; Tobler et al. 2008; Hsu et al. 2009; Takahashi et al. 2010; Wu et al. 2011; Berns and Bell 2012), within the parietal cortex (Berns et al. 2008), and/or within the anterior insula (Preuschoff et al. 2008; Bossaerts 2010; Mohr et al. 2010). Alternatively, (inverse) S-shaped probability weighting might constitute an emergent feature of the processing of probabilistic information by neurons (Gold and Shadlen 2007; Yang and Shadlen 2007; Soltani and Wang 2010; Pouget et al. 2013).

**The Probabilistic Reasoning Model**

Based on the findings documented in this chapter, a probabilistic reasoning (PR) model of the Bayesian brain is suggested, which basically reflects the tri-partitioned late positive complex. The conceptual outline of the PR model is given in Fig. 4.18, while Fig. 4.19 shows a formal outline. The PR model postulates the existence of a Bayesian reasoning unit (BRU) that interacts, in a reciprocal manner, with cognitive systems that process incoming environmental information (Haykin and Fuster 2014). The BRU computes, retains and updates two distinguishable probability distributions, one over the hidden state (beliefs; light gray color) and another one over the observable events (predictions; dark gray color). The PR model states that belief updating (Bayesian surprise (4.20)) and prediction updating (postdictive surprise (4.22)) are associated with trial-by-trial P3a and SW amplitude variations, respectively. The new predictions of the BRU set pre-adaptive biases on perceptual decisions, whereas BRU belief updating is based on the observation that results from these decisions. Note that predictive surprise ((4.23), related to trial-by-trial P3b amplitude variations) can be thought of as the magnitude of prediction errors induced by unpredicted or surprising

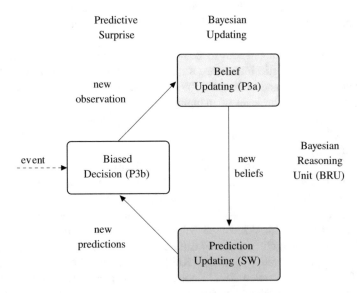

**Fig. 4.18** A conceptual outline of the probabilistic reasoning (PR) model of the tri-partitioned late positive complex. The PR model states that belief updating (Bayesian surprise (4.20)) and prediction updating (postdictive surprise (4.22)) are associated with trial-by-trial P3a and SW amplitude variations, respectively. The new predictions set pre-adaptive biases on perceptual decisions, whereas the belief updating is based on the new observation that results from these decisions. Predictive surprise (4.23) can be thought of as the magnitude of prediction errors induced by unpredicted or surprising events

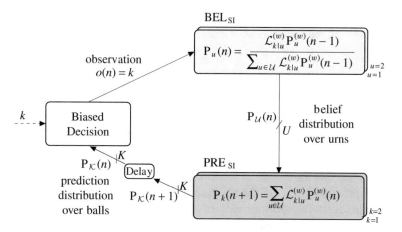

**Fig. 4.19** A formal outline of the PR model, in particular of the computational fine structure of the BRU. Units of time $(n-1, n)$ separate the dynamic evolution of beliefs over states (BEL$_{SI}$ (4.24)) that obeys Bayes' theorem (*lighter gray color*). Units of time $(n, n+1)$ also separate the dynamic evolution of predictions over observations (PRE$_{SI}$ (4.25)) as prescribed by Bayes' theorem (*darker gray color*)

events. Furthermore, the PR model states that the BRU is capable of Bayes-optimal updating: Firstly, it computes posterior distributions that take the prior and observation into account (belief updating, related to trial-by-trial P3a amplitude variations (4.20)). Secondly, prediction distributions for future observations are computed from posterior distributions (prediction updating, related to trial-by-trial SW amplitude variations (4.22)).

On the output branch of the BRU, predictions are used to set—preadaptive— biases on cognitive processing (Fuster 2014), whereas BRU belief updating is based on the incoming observation. Note that predictive surprise (related to trial-by-trial P3b amplitude variations) can be thought of as the magnitude of prediction errors induced by pre-adaptively biased processing within the cognitive processing stream. Predictive surprise could also be considered as the evolution of a decision variable, i.e., as the accumulation of evidence from bias levels to a decision threshold (Kopp 2008; O'Connell et al. 2012; Kelly and O'Connell 2013; Kira et al. 2015). Also, it remains unknown whether the P3a reflects the proper updating of beliefs, or an obligatory attentional process that forms part of the belief updating (which means an orienting response; Friedman et al. 2001; Barry and Rushby 2006; Kopp and Lange 2013).

The probabilistic reasoning model of the late positive complex can be regarded as a computational advancement of the most widely renowned and respected conceptual theory in the P300 field, i.e., the so-called "context updating" model (Donchin 1981; Donchin and Coles 1988). In short, this model postulates that the P300 is evoked in the service of meta-cognitive processes that are concerned with maintaining a proper representation of the environment, such as the mapping of probabilities on the environment, the deployment of attention, or the setting of priorities and biases.

**Next Steps**

The PR model suggests that the brain maintains and updates an internal model of the environment and weights prior beliefs and new observations proportionately within the BRU (Vilares and Körding 2011). But why did these neuronal connections evolve that are required for maintaining these computationally expensive internal models? According to the free-energy principle (which instantiates the Bayesian brain hypothesis), the driving force is the minimization of average predictive surprise (Friston 2010), and this minimization function is based on Bayes-optimal probabilistic inference, presumably modulated via the hyper-parameter $\zeta$. This line of reasoning suggests that meta-Bayesian learning of a policy for setting $\zeta$-values might occur in response to experienced predictive surprise values, thereby shifting the Bayesian observer towards higher or lower levels of uncertainty. Further research is required to investigate this proposition.

# Chapter 5
# Summary and Outlook

This work investigated how the human brain integrates sensory information to infer statistical regularities of the environment, and whether these processes are consistent with the Bayesian brain hypothesis. To this end, Bayes-optimal observer models have been proposed to predict trial-by-trial amplitude fluctuations of event-related potentials. A framework of Bayesian updating and predictive surprise was used to relate the observer models to neural activities, and Bayesian model selection has been employed to choose the best models. Two new observer models have proven to be superior: The DIF model, which combines the properties of two models taken from the literature, and the Bayesian observer model, which was directly derived from Bayes' theorem. By isolating discrete neural signatures of Bayesian updating and predictive surprise based on these observer models, fundamental evidence for the coding and computing of probability distributions in the human brain could be obtained. Thus, a network of brain areas could be suggested that may allow for Bayes-optimal probabilistic reasoning (Jaynes 1988; Dayan et al. 1995; Rangel et al. 2008; Bach and Dolan 2012; Summerfield and Tsetsos 2012). This evidence is consistent with, and strongly supports, the Bayesian brain hypothesis and, more specifically, the free-energy principle and theories of predictive coding. Such evidence based on experimental data was never reported before. These results were achieved in the following way:

In the first step, the DIF model was tested against models taken from the literature using trial-by-trial P300 amplitudes that were obtained at electrode Pz during performance of an oddball task. Conventional ERP analyses showed a clear effect of the stimulus probability and the preceding stimulus sequence on the P300 amplitudes, thus confirming two of the most consistently reported properties of the P300. The model-based analyses showed superiority of the DIF model over its competitors. While it was derived from models taken from the literature, it possesses important advantages over them: It is a completely formal model which does not incorporate descriptive factors or information which was not actually available to the subjects,

© Springer International Publishing Switzerland 2016
A. Kolossa, *Computational Modeling of Neural Activities for Statistical Inference*, DOI 10.1007/978-3-319-32285-8_5

but it can nevertheless account for sequential effects on P300 amplitudes. By using data from only one single ERP evidence for the coding of probability distributions in the brain could be collected.

In the second step, a Bayesian observer model and the urn-ball task, which equips the subjects with prior knowledge about all probabilistic conditions, were introduced. Two distinct probability distributions, the belief distribution over hidden states (BEL) and the prediction distribution over observable events (PRE), were derived from the Bayesian observer model. Data from four temporally and locally distinguishable ERPs (N250, P3a, P3b, SW) were used for model selection, in order to obtain evidence not only for the coding of probability distributions but for their updating as well. The results reveal that the N250 is best fitted by predictive surprise based on the DIF model with a uniform initial prior, while for the P3a the updating of beliefs about hidden states (Bayesian surprise based on the $BEL_{SI}$ distribution) proved to be superior. The P3b is best explained by predictive surprise based on the $PRE_{SI}$ distribution, and the SW is best fitted by the updating of predictions of observations (postdictive surprise based on the $PRE_{SI}$ distribution). These results provide evidence that the three components of the late positive complex represent distinct neural computations within a Bayesian observer who makes use of prior knowledge about the environment and employs inverse S-shaped probability weighting of the inference input. The N250 seems to reflect a faster, purely memory-based observer who cannot incorporate any prior knowledge. These findings suggest that the brain maintains different sophisticated models, and they are consistent with earlier reports (Dolan and Dayan 2013; Strauss et al. 2015).

Future work might benefit from this work in multiple ways. The urn-ball task proved very useful for studying Bayesian inference by the human brain. Future studies should rely on the urn-ball task to obtain ERP amplitudes for model-based analyses. The trial-by-trial ERP amplitude fluctuations of the dissociable components of the late positive complex are proven to be a valuable target for computational models of the neural bases of Bayesian inference. This warrants more attention to the late positive complex and the interaction of its components. Alterations to the urn-ball task can be envisaged which aim towards a better understanding of the discrimination between fast, memory-based and slow, model-based forms of surprise. Furthermore, allowing the hidden state to change during an episode of sampling and incorporating a non-unity state transition matrix into the observer model might yield further evidence for Bayes-optimal reasoning of the brain.

This work showed the importance of the usage of (inverse) S-shaped weighting of probabilities in attempts to model the neural bases of Bayesian inference. Future work should take the hereby documented fundamental effect of the probability weighting functions on Bayesian inference into account. It should investigate the effect of subject-individual values for the hyper-parameter $\zeta$ when applying probability weighting functions. Also, rules for the dynamical adaptation of $\zeta$ in dependence on the prediction error seem a promising approach for getting insights into higher levels of cognitive functions. Additionally, the approach for optimizing $\zeta$ based purely on the behavioral data offers a new approach to model optimization which avoids circularity and fully conforms to the theory that "... understanding cognitive functions reduces to assessing input-output relationships, where inputs are experimentally controlled stimuli and/or task instructions, and outputs are observed behavioural outcomes." (Rigoux and Daunizeau 2015).

The role of the signal-to-noise ratio for ERP research, which has only been a side issue in this work, needs to be systematically evaluated, and clear limits have to be shown. Even in current guidelines for EEG-based research (Keil et al. 2014) the importance of the signal-to-noise ratio is merely acknowledged, but neither quantitative descriptions nor limits are provided. Especially for model-based trial-by-trial analyses, guidelines stating the minimum number of trials per subject in each condition and the total number of subjects necessary for reliable statistical inference in dependence on the SNR should be derived. In the course of this, the effect of the correlation between the outputs of different observer models on the maximally achievable statistical strength of the results has to be investigated in light of the SNR. Furthermore, the SNR might actually prove as a meaningful, reproducible data feature for ERP latency estimation and data rejection. Therefore, the signal-to-noise ratio should without any doubt be a subject of future work.

This work relied on electrode-specific signals to derive the ERP. Due to volume conduction in the brain, these signals are the summed activities of the brain, and the interference of different sources impedes the analyses of single functions. Independent component analyses (Makeig et al. 2002) and source localization (Michel and Murray 2012) of the EEG signals may lead to the discovery of further aspects of Bayesian reasoning which are not dissociable in the summed activities obtained at a single electrode. Also, in light of the here documented interdependency of the components of the late positive complex, the information flow between the different sources (Delorme et al. 2011) offers the potential for collecting further evidence for the presence of the Bayesian reasoning unit and deeper insights into its functioning. Coupling the EEG data-based analyses with functional magnetic resonance imaging, which has, in contrast to the EEG, great spatial resolution at a poor temporal resolution, offers the potential for better identifying brain regions associated with cortical functions and discovering increasingly finer functional features (Debener et al. 2005; Rosa et al. 2010; Ullsperger and Debener 2010; Jorge et al. 2014).

It is the hope of this author that the evidence collected and the conclusions drawn in this work, along with the inspirations for future work, will contribute to make Lisman's (2015) assessment come true that "... history is likely to look back on the first half of the 21st century as the period during which the brain came to be understood."

# Appendix

## On the Equivalence of Count Functions and Digital Filters

The relation between the digital filtering approach of the DIF model and the count functions of the state-of-the-art models calls for some attention. First, it will be shown that the count functions (3.20) for $c_{S,k}(n)$ and (3.22) for $c_{L,k}(n)$ are equivalent to their recursive formulations (3.21) and (3.23), respectively. To this end the block diagram of Fig. 3.5 will be transformed to obtain the count functions and normalizing constants from the new resulting block diagrams.

Figure A.1a shows the block diagram of Fig. 3.5 according to (3.23) with some omitted notations. A delay of one trial is represented by $\boxed{T}$, the input signal is defined as $g_k(n)$ (3.19), the output signal as $c_{L,k}(n)$, and $(1-\gamma_{L,n-1})$ and $\gamma_{L,n-1}$ are time-dependent values. The multiplier $(1-\gamma_{L,n-1})$ can be moved to the right yielding the block diagram shown in Fig. A.1b. Figure A.1c shows the block diagram after moving $(1-\gamma_{L,n-1})$ even further to the right. Finally, when moving the multiplier $\frac{\gamma_{L,n-1}}{1-\gamma_{L,n-1}}$ to the right of the lower branch delay unit, the time dependency has to be accounted for as shown in Fig. A.1d. For the long-term filter an effective filter coefficient is obtained:

$$\gamma'_{L,n} = \gamma_{L,n}\frac{1-\gamma_{L,n-1}}{1-\gamma_{L,n}} \tag{A.1}$$

$$= e^{-\frac{1}{\beta_{L,n}}}\frac{1-e^{-\frac{1}{\beta_{L,n-1}}}}{1-e^{-\frac{1}{\beta_{L,n}}}},$$

while for the short-term filter this simplifies to

$$\gamma'_{S,n} = \gamma_S = e^{-\frac{1}{\beta_S}}. \tag{A.2}$$

© Springer International Publishing Switzerland 2016
A. Kolossa, *Computational Modeling of Neural Activities for Statistical Inference*, DOI 10.1007/978-3-319-32285-8

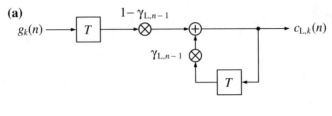

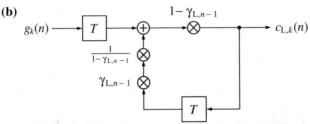

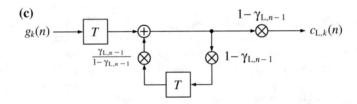

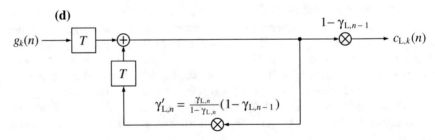

**Fig. A.1** Illustration of the equivalence of the digital filter (3.23) (Fig. 3.5) to the count function (3.22). **a** Block diagram of the DIF model's long-term memory filter of Fig. 3.5 and (3.23) with input signal $g_k(n)$ and output signal $c_{L,k}(n)$. **b** Block diagram of a filter equivalent to **a**, where the multiplier $(1 - \gamma_{L,n-1})$ has been moved to the right. **c** Block diagram of a filter equivalent to **b**, where the multiplier $(1 - \gamma_{L,n-1})$ has been moved even further to the right. **d** Block diagram of a filter equivalent to **c**, where the multiplier $\frac{\gamma_{L,n-1}}{1 - \gamma_{L,n-1}}$ has been moved to the right of the delay unit in the lower branch

The time-varying discrete-time impulse response of the long-term filter depicted in Figure A.1d is given as

$$h_{L,v}(n) = \varepsilon(n - v - 1) \frac{1 - \gamma_{L,n-1}}{\gamma'_{L,n}} \prod_{v=v+1}^{n} \gamma'_{L,v} \qquad (A.3)$$

with $n$ ($> \nu$) being the currently observed trial, and $\nu$ being the trial when the stimulus initiated the impulse response. We make use of the step function

$$\varepsilon(n - \nu - 1) = \begin{cases} 0, & n - \nu - 1 < 0 \\ 1, & n - \nu - 1 \geq 0. \end{cases} \tag{A.4}$$

Substituting (A.1) in (A.3) yields

$$h_{L,\nu}(n) = \varepsilon(n - \nu - 1) \frac{1 - \gamma_{L,n}}{\gamma_{L,n}} \prod_{\upsilon=\nu+1}^{n} \gamma'_{L,\upsilon}$$

$$= \varepsilon(n - \nu - 1) \frac{1}{C_{L,n}} \prod_{\upsilon=\nu+1}^{n} \gamma'_{L,\upsilon}, \tag{A.5}$$

with the dynamic normalizing value $C_{L,n} = \frac{\gamma_{L,n}}{1 - \gamma_{L,n}}$. The input–output relation of this time-varying linear system is given by (c.f. Claasen and Mecklenbräuker 1982; Prater and Loeffler 1992)

$$c_{L,k}(n) = \sum_{\nu=-\infty}^{\infty} h_{L,\nu}(n) g_k(\nu). \tag{A.6}$$

Due to the step function (A.4) in the impulse response (A.3) only stimuli at trials $\nu \leq n - 1$ contribute to the output at trial $n$, and

$$c_{L,k}(n) = \sum_{\nu=-\infty}^{n-1} h_{L,\nu}(n) g_k(\nu)$$

$$= \frac{1}{C_{L,n}} \sum_{\nu=-\infty}^{n-1} \left( \prod_{\upsilon=\nu+1}^{n} \gamma'_{L,\upsilon} \right) g_k(\nu)$$

$$= \frac{1}{C_{L,n}} \sum_{\nu=-\infty}^{n-1} \gamma_{L,n}(\nu) \, g_k(\nu) \tag{A.7}$$

is obtained, which is exactly the long-term count function (3.22) with $\gamma_{L,n}(\nu) = \prod_{\upsilon=\nu+1}^{n} \gamma'_{L,\upsilon} = \prod_{\upsilon=\nu+1}^{n} \gamma_{L,\upsilon} \frac{1-\gamma_{L,\upsilon-1}}{1-\gamma_{L,\upsilon}}$, and $g_k(\nu)$, as before. Note that for the short-term filter $\prod_{\upsilon=\nu+1}^{n} \gamma_S = \gamma_S^{n-\nu}$ and $C_S = \frac{\gamma_S}{1-\gamma_S}$, which simplifies (A.7) to the discrete-time convolution

$$c_{S,k}(n) = \frac{1}{C_S} \sum_{\nu=-\infty}^{n-1} \gamma_S^{n-\nu} g_k(\nu), \tag{A.8}$$

which is the short-term count function (3.20). Thus, equivalence has been shown between the count functions (3.20) and (3.22), and their recursive formulations (3.21) and (3.23), respectively.

# Bibliography

Achtziger A, Alós-Ferrer C, Hügelschäfer S, Steinhauser M (2014) The neural basis of belief updating and rational decision making. Soc Cogn Affect Neurosci 9:55–62

Alba JW, Chromiak W, Hasher L, Attig MS (1980) Automatic encoding of category size information. J Exp Psychol Hum Learn Mem 6:370–378

Bach DR (2014) Das Bayesianische Gehirn. Gehirn und Geist 1–2:54–59

Bach DR, Dolan RJ (2012) Knowing how much you don't know: a neural organization of uncertainty estimates. Nat Rev Neurosci 13:572–586

Baddeley A (2003) Working memory: looking back and looking forward. Nat Rev Neurosci 4:829–839

Baldi P, Itti L (2010) Of bits and wows: a Bayesian theory of surprise with applications to attention. Neural Netw 23:649–666

Barceló F, Periáñez JA, Nyhus E (2008) An information theoretical approach to task-switching: evidence from cognitive brain potentials in humans. Front Hum Neurosci 1:13

Barlow H (2001) Redundancy reduction revisited. Netw Comput Neural Syst 12:241–253

Barnard GA (1949) Statistical inference. J Roy Stat Soc Ser B 11:115–149

Barrett LF (2009) Understanding the mind by measuring the brain: lessons from measuring behavior (commentary on Vul et al, 2009). Perspect Psychol Sci 4:314–318

Barry RJ, Rushby JA (2006) An orienting reflex perspective on anteriorisation of the P3 of the event-related potential. Exp Brain Res 173:539–545

Başar E (1980) EEG-brain Dynamics: Relation Between EEG and Brain Evoked Potentials. Elsevier-North-Holland Biomedical Press, Amsterdam, NL

Beal MJ, Ghahramani Z (2003) The variational Bayesian EM algorithm for incomplete data: with application to scoring graphical model structures. In: Bernado JM, Bayarri MJ, Berger JO, Dawid AP, Heckerman D, Smith AFM, West M (eds) Bayesian Statistics 7, vol 7. Oxford University Press, Oxford, UK, pp 453–464

Beal MJ (2003) Variational algorithms for approximate Bayesian inference. Ph.D. thesis, University College London, UK

Beim Graben P (2001) Estimating and improving the signal-to-noise ratio of time series by symbolic dynamics. Phys Rev E 64:051104-1–051104-15

Berger H (1929) Über das Elektrenkephalogramm des Menschen. Archiv für Psychiatrie und Nervenkrankheiten 87:527–570

Berns GS, Bell E (2012) Striatal topography of probability and magnitude information for decisions under uncertainty. NeuroImage 59:3166–3172

Berns GS, Capra CM, Chappelow J, Moore S, Noussair C (2008) Nonlinear neurobiological probability weighting functions for aversive outcomes. NeuroImage 39:2047–2057

Biau DJ, Jolles BM, Porcher R (2010) P value and the theory of hypothesis testing: an explanation for new researchers. Clin Orthop Relat Res 468:885–892

Blankertz B, Curio G, Müller KR (2002) Classifying single trial EEG: towards brain computer interfacing. Adv Neural Inf Process Syst 1:157–164

Bossaerts P (2010) Risk and risk prediction error signals in anterior insula. Brain Struct Funct 214:645–653

Carron AV, Bailey DA (1969) Evidence for reliable individual differences in intra-individual variability. Percep Mot Skills 28:843–846

Cavagnaro DR, Pitt MA, Gonzalez R, Myung JI (2013) Discriminating among probability weighting functions using adaptive design optimization. J Risk Uncertain 47:255–289

Claasen TACM, Mecklenbräuker WFG (1982) On stationary linear time-varying systems. IEEE Trans Circ Syst 29:169–184

Clark A (2013) Whatever next? Predictive brains, situated agents, and the future of cognitive science. Behav Brain Sci 36:181–253

Cohen J (1994) The earth is round ($p < .05$). Am Psychol 49:997–1003

Coppola R, Tabor R, Buchsbaum MS (1978) Signal to noise ratio and response variability measurements in single trial evoked potentials. Electroencephalogr Clin Neurophysiol 44:214–222

Cowan N (2005) Working Memory Capacity. Psychology Press, New York, NY

Czanner G, Sarma SV, Ba D, Eden UT, Wu W, Eskandar E, Lim HH, Temereanca S, Suzuki WA, Brown EN (2015) Measuring the signal-to-noise ratio of a neuron. Proc Natl Acad Sci 112:7141–7146

da Silva FL (2013) EEG and MEG: relevance to neuroscience. Neuron 80:1112–1128

d'Acremont M, Schultz W, Bossaerts P (2013) The human brain encodes event frequencies while forming subjective beliefs. J Neurosci 33:10887–10897

Daunizeau J, Den Ouden HEM, Pessiglione M, Kiebel SJ, Friston KJ, Stephan KE (2010) Observing the observer (II): deciding when to decide. PLOS One 5:e15555

Daunizeau J, Adam V, Rigoux L (2014) VBA: a probabilistic treatment of nonlinear models for neurobiological and behavioural data. PLOS Comput Biol 10:e1003441

Dayan P, Hinton GE, Neal RM, Zemel RS (1995) The Helmholtz machine. Neural Comput 7:889–904

de Lange FP, Jensen O, Dehaene S (2010) Accumulation of evidence during sequential decision making: the importance of top-down factors. J Neurosci 30:731–738

Debener S, Ullsperger M, Siegel M, Fiehler K, von Cramon DY, Engel AK (2005) Trial-by-trial coupling of concurrent electroencephalogram and functional magnetic resonance imaging identifies the dynamics of performance monitoring. J Neurosci 25:11730–11737

Delorme A, Makeig S (2004) EEGLAB: an open source toolbox for analysis of single-trial EEG dynamics including independent component analysis. J Neurosci Methods 134:9–21

Delorme A, Mullen T, Kothe C, Acar ZA, Bigdely-Shamlo N, Vankov A, Makeig S (2011) EEGLAB, SIFT, NFT, BCILAB, and ERICA: new tools for advanced EEG processing. Comput Intell Neurosci 2011:10

Dempster AP, Rubin DB, Tsutakawa RK (1981) Estimation in covariance components models. J Am Stat Assoc 76:341–353

Diaconescu AO, Mathys C, Weber LAE, Daunizeau J, Kasper L, Lomakina EI, Fehr E, Stephan KE (2014) Inferring on the intentions of others by hierarchical Bayesian learning. PLOS Comput Biol 10:e1003810

Dien J, Spencer KM, Donchin E (2004) Parsing the late positive complex: mental chronometry and the ERP components that inhabit the neighborhood of the P300. Psychophysiology 41:665–678

Dolan RJ, Dayan P (2013) Goals and habits in the brain. Neuron 80:312–325

Donchin E (1981) Surprise! Surprise? Psychophysiology 18:493–513

Donchin E, Coles MG (1988) Is the P300 component a manifestation of context updating? Behav Brain Sci 11:357–427

Doya K, Ishii S, Pouget A, Rao RPN (2007) Bayesian brain: probabilistic approaches to neural coding. MIT Press, Cambridge, MA

Duncan-Johnson CC, Donchin E (1977) On quantifying surprise: the variation of event-related potentials with subjective probability. Psychophysiology 14:456–467

Egner T, Monti JM, Summerfield C (2010) Expectation and surprise determine neural population responses in the ventral visual stream. J Neurosci 30:16601–16608

Fahrmeir L, Tutz G (1994) Multivariate statistical modelling based on generalized linear models. Springer, New York, NY

Fennell J, Baddeley R (2012) Uncertainty plus prior equals rational bias: an intuitive Bayesian probability weighting function. Psychol Rev 119:878–887

Fiedler K (2011) Voodoo correlations are everywhere-not only in neuroscience. Perspect Psychol Sci 6:163–171

Fisher RA (1926) The arrangement of field experiments. J Minist Agric 33:503–513

FitzGerald THB, Schwartenbeck P, Moutoussis M, Dolan RJ, Friston KJ (2015) Active inference, evidence accumulation, and the urn task. Neural Comput 27:306–328

Folstein JR, Van Petten C (2008) Influence of cognitive control and mismatch on the N2 component of the ERP: a review. Psychophysiology 45:152–170

Ford JM, Roach BJ, Miller RM, Duncan CC, Hoffman RE, Mathalon DH (2010) When it's time for a change: failures to track context in schizophrenia. Int J Psychophysiol 78:3–13

Fox CR, Poldrack RA (2009) Prospect theory and the Brain. In: Glimcher PW, Camerer CF, Fehr E, Poldrack RA (eds) Neuroeconomics: decision making and the Brain. Elsevier, London, UK, pp 145–173

Friedman D, Cycowicz YM, Gaeta H (2001) The novelty P3: an event-related brain potential (ERP) sign of the brain's evaluation of novelty. Neurosci Biobehav Rev 25:355–373

Friston KJ (2002) Beyond phrenology: what can neuroimaging tell us about distributed circuitry? Ann Rev Neurosci 25:221–250

Friston KJ (2005) A theory of cortical responses. Phil Trans R Soc B Biol Sci 360:815–836

Friston KJ (2012) Ten ironic rules for non-statistical reviewers. NeuroImage 61:1300–1310

Friston KJ, Glaser DE, Henson RNA, Kiebel SJ, Phillips C, Ashburner J (2002) Classical and Bayesian inference in neuroimaging: applications. NeuroImage 16:484–512

Friston KJ, Mattout J, Trujillo-Bareto N, Ashburner J, Penny WD (2007) Variational free energy and the Laplace approximation. NeuroImage 34:220–234

Friston KJ, Penny WD (2003) Classical and Bayesian inference. In: Frackowiak RSJ, Friston KJ, Frith C, Dolan R, Price CJ, Zeki S, Ashburner J, Penny WD (eds) Human Brain function. Academic Press, London, UK, pp 911–970

Friston KJ, Penny WD, Phillips C, Kiebel SJ, Hinton G, Ashburner J (2002) Classical and Bayesian inference in neuroimaging: theory. NeuroImage 16:465–483

Friston KJ (2010) The free-energy principle: a unified brain theory? Nat Rev Neurosci 11:127–138

Furl N, Averbeck BB (2011) Parietal cortex and insula relate to evidence seeking relevant to reward-related decisions. J Neurosci 31:17572–17582

Fuster JM (2014) The prefrontal cortex makes the brain a preadaptive system. Proc IEEE 102:417–426

García-Larrea L, Cézanne-Bert G (1998) P3, positive slow wave and working memory load: a study on the functional correlates of slow wave activity. Electroencephalogr Clin Neurophysiol/Evoked Potentials Sect 108:260–273

Garrido MI, Kilner JM, Kiebel SJ, Friston KJ (2009) Dynamic causal modeling of the response to frequency deviants. J Neurophysiol 101:2620–2631

Gold JI, Shadlen MN (2007) The neural basis of decision making. Ann Rev Neurosci 30:535–574

Gonsalvez CJ, Polich J (2002) P300 amplitude is determined by target-to-target interval. Psychophysiology 39:388–396

Gonzalez R, Wu G (1999) On the shape of the probability weighting function. Cogn Psychol 38:129–166

Goodman SN (1999a) Toward evidence-based medical statistics. 1: the P value fallacy. Ann Intern Med 130:995–1004

Goodman SN (1999b) Toward evidence-based medical statistics. 2: the Bayes factor. Ann Intern Med 130:1005–1013

Gratton G, Coles MGH, Donchin E (1983) A new method for off-line removal of ocular artifact. Electroencephalogr Clin Neurophysiol 55:468–484

Grether DM (1980) Bayes rule as a descriptive model: the representativeness heuristic. Q J Econ 95:537–557

Grether DM (1992) Testing Bayes rule and the representativeness heuristic: some experimental evidence. J Econ Behav Organ 17:31–57

Hampton AN, Bossaerts P, O'Doherty JP (2006) The role of the ventromedial prefrontal cortex in abstract state-based inference during decision making in humans. J Neurosci 26:8360–8367

Harrison LM, Bestmann S, Rosa MJ, Penny W, Green GGR (2011) Time scales of representation in the human brain: weighing past information to predict future events. Front Hum Neurosci 5:37

Hasher L, Zacks RT (1984) Automatic processing of fundamental information: the case of frequency of occurrence. Am Psychol 39:1372–1388

Haykin S, Fuster JM (2014) On cognitive dynamic systems: cognitive neuroscience and engineering learning from each other. Proc IEEE 102:608–628

Hilbert M (2012) Toward a synthesis of cognitive biases: how noisy information processing can bias human decision making. Psychol Bull 138:211–237

Hillyard SA, Picton TW (1987) Electrophysiology of cognition. In: Plum F (ed) Handbook of physiology. American Physiological Society, Bethesda, MD, pp 519–584

Hintzman DL (1976) Repetition and memory. In: Bower GH (ed) The psychology of learning and motivation, vol 10. Academic Press, New York, NY, pp 47–91

Hoeting JA, Madigan D, Raftery AE, Volinsky CT (1999) Bayesian model averaging: a tutorial. Stat Sci 14:382–401

Hoijtink H (2012) Informative hypotheses: theory and practice for behavioral and social scientists. CRC Press, New York, NY

Hoijtink H, Klugkist I, Boelen PA (2008) Bayesian evaluation of informative hypotheses. Springer, New York, NY

Holmes AP, Friston KJ (1998) Generalisability, random effects & population inference. NeuroImage 7:754

Hsu M, Krajbich I, Zhao C, Camerer CF (2009) Neural response to reward anticipation under risk is nonlinear in probabilities. J Neurosci 29:2231–2237

Ingber L (1996) Adaptive simulated annealing (ASA): lessons learned. Control Cybern 25:33–54

Itti L, Baldi P (2009) Bayesian surprise attracts human attention. Vis. Res 49:1295–1306

ITU (2006a) Mean opinion score (MOS) terminology. ITU-T Recommendation P.800.1

ITU (2006b) Vocabulary for performance and quality of service. ITU-T Recommendation P.10/G.100

ITU (2007) Wideband extension to recommendation P. 862 for the assessment of wideband telephone networks and speech codecs. ITU-T Recommendation P. 862.2

Jaynes ET (1988) How does the brain do plausible reasoning? In: Erickson GJ, Smith CR (eds) Maximum-entropy and Bayesian methods in science and engineering, vol 1. Kluwer Academic PublishersDordrecht, NL, pp 1–24

Jaynes ET (2003) Probability theory: the logic of science. Cambridge University Press, Cambridge, UK

Jentzsch I, Sommer W (2001) Sequence-sensitive subcomponents of P300: topographical analyses and dipole source localization. Psychophysiology 38:607–621

Jonides J, Lewis RL, Nee DE, Lustig CA, Berman MG, Moore KS (2008) The mind and brain of short-term memory. Ann Rev Psychol 59:193–224

Jorge J, Van Der Zwaag W, Figueiredo P (2014) EEG-fMRI integration for the study of human brain function. NeuroImage 102:24–34

Kahneman D, Tversky A (1979) Prospect theory: an analysis of decision under risk. Econometrica 47:263–291

Kass RE, Raftery AE (1995) Bayes factors. J Am Stat Assoc 90:773–795

Keil A, Debener S, Gratton G, Junghöfer M, Kappenman ES, Luck SJ, Luu P, Miller GA, Yee CM (2014) Committee report: publication guidelines and recommendations for studies using electroencephalography and magnetoencephalography. Psychophysiology 51:1–21

Kelly SP, O'Connell RG (2013) Internal and external influences on the rate of sensory evidence accumulation in the human brain. J Neurosci 33:19434–19441

Kersten D, Mamassian P, Yuille A (2004) Object perception as Bayesian inference. Annl Rev Psychol 55:271–304

Kiebel S, Holmes AP (2003) The general linear model. In: Frackowiak RSJ, Friston KJ, Frith C, Dolan R, Price CJ, Zeki S, Ashburner J, Penny WD (eds) Human Brain function. Academic Press, London, UK, pp 725–760

Kira S, Yang T, Shadlen MN (2015) A neural implementation of Wald's sequential probability ratio test. Neuron 85:861–873

Kleinbaum D, Kupper L, Nizam A, Rosenberg E (2013) Applied regression analysis and other multivariable methods, 5th edn. Cengage Learning, Boston, MA

Knill DC, Pouget A (2004) The Bayesian brain: the role of uncertainty in neural coding and computation for perception and action. Trends Neurosci 27:712–719

Kok P, Failing MF, de Lange FP (2014) Prior expectations evoke stimulus templates in the primary visual cortex. J Cogn Neurosci 26:1546–1554

Kolda TG, Lewis RM, Torczon V (2006) A generating set direct search augmented Lagrangian algorithm for optimization with a combination of general and linear constraints. Sandia National Laboratories Report, Albuquerque, NM

Kolossa A, Abel J, Fingscheidt T (2016) Comparing instrumental measures of speech quality using Bayesian model selection: correlations can be misleading! In: Proceedings of ICASSP, 2016. Shanghai, China, pp 634–638

Kolossa A, Fingscheidt T, Wessel K, Kopp B (2013) A model-based approach to trial-by-trial P300 amplitude fluctuations. Front Hum Neurosci 6:359

Kolossa A, Kopp B, Fingscheidt T (2015) A computational analysis of the neural bases of Bayesian inference. NeuroImage 106:222–237

Kopp B (2008) The P300 component of the event-related brain potential and Bayes' theorem. In: Sun MK (ed) Cognitive sciences at the leading edge. Nova Science Publishers, New York, NY, pp 87–96

Kopp B, Lange F (2013) Electrophysiological indicators of surprise and entropy in dynamic task-switching environments. Front Hum Neurosci 7:300

Kriegeskorte N, Lindquist MA, Nichols TE, Poldrack RA, Vul E (2010) Everything you never wanted to know about circular analysis, but were afraid to ask. J Cereb Blood Flow Metab 30:1551–1557

Kriegeskorte N, Simmons WK, Bellgowan PSF, Baker CI (2009) Circular analysis in systems neuroscience: the dangers of double dipping. Nat Neurosci 12:535–540

Lazar NADiscussion of puzzlingly high correlations in fMRI studies of emotion, personality, and social cognition by Vul, et al (2009) Perspect Psychol Sci 4:308–309

Lee D, Rushworth MFS, Walton ME, Watanabe M, Sakagami M (2007) Functional specialization of the primate frontal cortex during decision making. J Neurosci 27:8170–8173

Leuthold H, Sommer W (1993) Stimulus presentation rate dissociates sequential effects in event-related potentials and reaction times. Psychophysiology 30:510–517

Lieberman MD, Berkman ET, Wager TDCorrelations in social neuroscience aren't voodoo: commentary on Vul, et al (2009) Perspect Psychol Sci 4:299–307

Lieder F, Daunizeau J, Garrido MI, Friston KJ, Stephan KE (2013) Modelling trial-by-trial changes in the mismatch negativity. PLOS Comput Biol 9:e1002911

Lindquist MA, Gelman A (2009) Correlations and multiple comparisons in functional imaging: a statistical perspective (commentary on Vul et al., 2009). Perspect Psychol Sci 4:310–313

Lisman J (2015) The challenge of understanding the brain: where we stand in 2015. Neuron 86:864–882

Lu ZL, Neuse J, Madigan S, Dosher BA (2005) Fast decay of iconic memory in observers with mild cognitive impairments. Proc Natl Acad Sci USA 102:1797–1802

Luck SJ (2004) Ten simple rules for designing and interpreting ERP experiments. In: Handy TC (ed) Event-related potentials: a methods handbook. MIT Press, Cambridge, MA, pp 17–32

Luck SJ (2014) An introduction to the event-related potential technique. MIT Press, Cambridge, MA

MacKay DJC (1992) Bayesian interpolation. Neural Comput 4:415–447

MacKay DJC (2003) Information theory, inference, and learning algorithms. Cambridge University Press, Cambridge, UK

Makeig S, Westerfield M, Jung TP, Enghoff S, Townsend J, Courchesne E, Sejnowski TJ (2002) Dynamic brain sources of visual evoked responses. Science 295:690–694

Makeig S, Debener S, Onton J, Delorme A (2004) Mining event-related brain dynamics. Trends Cogn Sci 8:204–210

Mars RB, Debener S, Gladwin TE, Harrison LM, Haggard P, Rothwell JC, Bestmann S (2008) Trial-by-trial fluctuations in the event-related electroencephalogram reflect dynamic changes in the degree of surprise. J Neurosci 28:12539–12545

Mars RB, Shea NJ, Kolling N, Rushworth MFS (2012) Model-based analyses: promises, pitfalls, and example applications to the study of cognitive control. Q J Exp Psychol 65:252–267

Masicampo EJ, Lalande DR (2012) A peculiar prevalence of p values just below.05. Q J Exp Psychol 65:2271–2279

Matsuda I, Nittono H (2015) Motivational significance and cognitive effort elicit different late positive potentials. Clin Neurophysiol 126:304–313

Michel CM, Murray MM (2012) Towards the utilization of EEG as a brain imaging tool. NeuroImage 61:371–385

Möcks J, Gasser T, Köhler W (1988) Basic statistical parameters of event-related potentials. J Psychophysiol 2:61–70

Mohr PN, Biele G, Heekeren HR (2010) Neural processing of risk. J Neurosci 30:6613–6619

Myung IJ, Pitt MA (1997) Applying Occam's razor in modeling cognition: a Bayesian approach. Psychon Bull Rev 4:79–95

Neal RM, Hinton GE (1998) A view of the EM algorithm that justifies incremental, sparse, and other variants. In: Jordan MI (ed) Learning in graphical models. Kluwer Academic Publishers, Dordecht, NL, pp 355–368

Neyman J, Pearson ES (1933) On the problem of the most efficient tests of statistical hypotheses. Phil Trans R Soc Lond Math Phys Eng Sci 231:289–337

O'Connell RG, Dockree PM, Kelly SP (2012) A supramodal accumulation-to-bound signal that determines perceptual decisions in humans. Nat Neurosci 15:1729–1735

Oldfield RC (1971) The assessment and analysis of handedness: the Edinburgh inventory. Neuropsychologia 9:97–113

O'Reilly JX, Schüffelgen U, Cuell SF, Behrens TEJ, Mars RB, Rushworth MFS (2013) Dissociable effects of surprise and model update in parietal and anterior cingulate cortex. Proc Natl Acad Sci 110:E3660–E3669

Osborne JW (2010) Challenges for quantitative psychology and measurement in the 21st century. Front Psychol 1:1

Ostwald D, Spitzer B, Guggenmos M, Schmidt TT, Kiebel SJ, Blankenburg F (2012) Evidence for neural encoding of Bayesian surprise in human somatosensation. NeuroImage 62:177–188

Paukkunen AKO, Leminen MM, Sepponen R (2010) Development of a method to compensate for signal quality variations in repeated auditory event-related potential recordings. Front Neuroeng 3:2

Penny WD (2012) Comparing dynamic causal models using AIC, BIC and free energy. NeuroImage 59:319–330

Penny WD, Holmes AP, Friston KJ (2003) Hierarchical models. In: Frackowiak RSJ, Friston KJ, Frith C, Dolan R, Price CJ, Zeki S, Ashburner J, Penny WD (eds) Human Brain function. Academic Press

Penny WD, Stephan KE, Mechelli A, Friston KJ (2004) Comparing dynamic causal models. NeuroImage 22:1157–1172

Penny WD, Stephan KE, Daunizeau J, Rosa MJ, Friston KJ, Schofield TM, Leff AP (2010) Comparing families of dynamic causal models. PLOS Comput Biol 6:e1000709

Phillips LD, Edwards W (1966) Conservatism in a simple probability inference task. J Exp Psychol 72:346–354

Picton TW, Bentin S, Berg P, Donchin E, Hillyard SA, Johnson R Jr, Miller GA, Ritter W, Ruchkin DS, Rugg MD et al (2000) Guidelines for using human event-related potentials to study cognition: recording standards and publication criteria. Psychophysiology 37:127–152

Pitt MA, Myung IJ (2002) When a good fit can be bad. Trends Cogn Sci 6:421–425

Polich J (2007) Updating P300: an integrative theory of P3a and P3b. Clin Neurophysiol 118:2128–2148

Pouget A, Beck JM, Ma WJ, Latham PE (2013) Probabilistic brains: knowns and unknowns. Nat Neurosci 16:1170–1178

Prater JS, Loeffler CM (1992) Analysis and design of periodically time-varying IIR filters, with applications to transmultiplexing. IEEE Trans Signal Process 40:2715–2725

Prelec D (1998) The probability weighting function. Econometrica 66:497–527

Preuschoff K, Bossaerts P, Quartz SR (2006) Neural differentiation of expected reward and risk in human subcortical structures. Neuron 51:381–390

Preuschoff K, Quartz SR, Bossaerts P (2008) Human insula activation reflects risk prediction errors as well as risk. J Neurosci 28:2745–2752

Puce A, Berkovic SF, Cadusch PJ, Bladin PF (1994) P3 latency jitter assessed using 2 techniques. I. Simulated data and surface recordings in normal subjects. Electroencephalogr Clin Neurophysiol 92:352–364

Rangel A, Camerer C, Montague PR (2008) A framework for studying the neurobiology of value-based decision making. Nat Rev Neurosci 9:545–556

Rauss K, Schwartz S, Pourtois G (2011) Top-down effects on early visual processing in humans: a predictive coding framework. Neurosci Biobehav Rev 35:1237–1253

Raz J, Turetsky B, Fein G (1988) Confidence intervals for the signal-to-noise ratio when a signal embedded in noise is observed over repeated trials. IEEE Trans Biomed Eng 35:646–649

Rigoux L, Daunizeau J (2015) Dynamic causal modelling of brain-behaviour relationships. NeuroImage 117:202–221

Ritter W, Vaughan HG (1969) Averaged evoked responses in vigilance and discrimination: a reassessment. Science 164:326–328

Rix AW, Beerends JG, Hollier MP, Hekstra AP (2001) Perceptual evaluation of speech quality (PESQ)–A new method for speech quality assessment of telephone networks and codecs. Proceedings of ICASSP, 2001, vol 2. Salt Lake City, Utah, pp 749–752

Robert C (2007) The Bayesian choice: from decision-theoretic foundations to computational implementation. Springer, New York, NY

Rosa MJ, Daunizeau J, Friston KJ (2010) EEG-fMRI integration: a critical review of biophysical modeling and data analysis approaches. J Integr Neurosci 9:453–476

Rosnow RL, Rosenthal R (1989) Statistical procedures and the justification of knowledge in psychological science. Am Psychol 44:1276–1284

Royall R (1997) Statistical evidence: a likelihood paradigm. Chapman & Hall/CRC, New York

Ruchkin DS, Johnson R, Mahaffey D, Sutton S (1988) Toward a functional categorization of slow waves. Psychophysiology 25:339–353

Rutherford A (2001) Introducing ANOVA and ANCOVA: a GLM approach. Sage, London

Schimmel H, Rapin I, Cohen MM (1974) Improving evoked response audiometry with special reference to the use of machine scoring. Int J Audiol 13:33–65

Schimmel H (1967) The ($\pm$) reference: accuracy of estimated mean components in average response studies. Science 157:92–94

Sedlmeier PE, Betsch TE (2002) ETC. Oxford University Press, New York, NY, Frequency processing and cognition

Shannon CE, Weaver W (1948) The mathematical theory of communication. Commun Bell Syst Tech J 27:379–423

Soltani A, Wang XJ (2010) Synaptic computation underlying probabilistic inference. Nat Neurosci 13:112–119

Spencer KM, Dien J, Donchin E (2001) Spatiotemporal analysis of the late ERP responses to deviant stimuli. Psychophysiology 38:343–358

Spratling MW (2010) Predictive coding as a model of response properties in cortical area V1. J Neurosci 30:3531–3543

Squires KC, Wickens C, Squires NK, Donchin E (1976) The effect of stimulus sequence on the waveform of the cortical event-related potential. Science 193:1142–1146

Stephan KE, Penny WD, Daunizeau J, Moran RJ, Friston KJ (2009) Bayesian model selection for group studies. NeuroImage 46:1004–1017

Stephan KE, Weiskopf N, Drysdale PM, Robinson PA, Friston KJ (2007) Comparing hemodynamic models with DCM. NeuroImage 38:387–401

Strange BA, Duggins A, Penny WD, Dolan RJ, Friston KJ (2005) Information theory, novelty and hippocampal responses: unpredicted or unpredictable? Neural Netw 18:225–230

Strauss M, Sitt JD, King JR, Elbaz M, Azizi L, Buiatti M, Naccache L, van Wassenhove V, Dehaene S (2015) Disruption of hierarchical predictive coding during sleep. Proc Nat Acad Sci 112:E1353–E1362

Summerfield C, Egner T (2009) Expectation (and attention) in visual cognition. Trends Cogn Sci 13:403–409

Summerfield C, Egner T, Greene M, Koechlin E, Mangels J, Hirsch J (2006) Predictive codes for forthcoming perception in the frontal cortex. Science 314:1311–1314

Summerfield C, Tsetsos K (2012) Building bridges between perceptual and economic decision-making: neural and computational mechanisms. Front Neurosci 6:70

Sutton S, Braren M, Zubin J, John ER (1965) Evoked-potential correlates of stimulus uncertainty. Science 150:1187–1188

Sutton S, Ruchkin DS (1984) The late positive complex. Ann N Y Acad Sci 425:1–23

Takahashi H, Matsui H, Camerer C, Takano H, Kodaka F, Ideno T, Okubo S, Takemura K, Arakawa R, Eguchi Y et al (2010) Dopamine D1 receptors and nonlinear probability weighting in risky choice. J Neurosci 30:16567–16572

Tobler PN, Christopoulos GI, O'Doherty JP, Dolan RJ, Schultz W (2008) Neuronal distortions of reward probability without choice. J Neurosci 28:11703–11711

Towey J, Rist F, Hakerem G, Ruchkin DS, Sutton S (1980) N250 latency and decision time. Bull Psychon Soc 15:365–368

Trepel C, Fox CR, Poldrack RA (2005) Prospect theory on the brain? Toward a cognitive neuroscience of decision under risk. Cogn Brain Res 23:34–50

Tversky A, Kahneman D (1992) Advances in prospect theory: cumulative representation of uncertainty. J Risk Uncertainty 5:297–323

Ullsperger M, Debener S (2010) Simultaneous EEG and fMRI: recording, analysis, and application. Oxford University Press, New York, NY

Underwood BJ (1969) Attributes of memory. Psychol Rev 76:559–573

Van de Schoot R, Hoijtink H, Romeijn JW (2011) Moving beyond traditional null hypothesis testing: evaluating expectations directly. Front Psychol 2:24

Varey CA, Mellers BA, Birnbaum MH (1990) Judgments of proportions. J Exp Psychol Hum Percept Perform 16:613–625

Vilares I, Körding K (2011) Bayesian models: the structure of the world, uncertainty, behavior, and the brain. Ann N Y Acad Sci 1224:22–39

Vilares I, Howard JD, Fernandes HL, Gottfried JA, Körding KP (2012) Differential representations of prior and likelihood uncertainty in the human brain. Curr Biol 22:1641–1648

Viola FC, Thorne JD, Bleeck S, Eyles J, Debener S (2011) Uncovering auditory evoked potentials from cochlear implant users with independent component analysis. Psychophysiology 48:1470–1480

Vul E, Harris C, Winkielman P, Pashler H (2009) Puzzlingly high correlations in fMRI studies of emotion, personality, and social cognition. Perspect Psychol Sci 4:274–290

Vul E, Pashler H (2012) Voodoo and circularity errors. NeuroImage 62:945–948

Wagenmakers EJ (2007) A practical solution to the pervasive problems of p values. Psychon Bull Rev 14:779–804

Wainer H (1999) One cheer for null hypothesis significance testing. Psychol Methods 4:212–213

Weiss DJ (2006) Analysis of variance and functional measurement: a practical guide. Oxford University Press, New York, NY

Winkler I, Czigler I (2011) Evidence from auditory and visual event-related potential (ERP) studies of deviance detection (MMN and vMMN) linking predictive coding theories and perceptual object representations. Int J Psychophysiol 83:132–143

Woolrich MW (2012) Bayesian inference in fMRI. NeuroImage 62:801–810

Wu SW, Delgado MR, Maloney LT (2011) The neural correlates of subjective utility of monetary outcome and probability weight in economic and in motor decision under risk. J Neurosci 31:8822–8831

Yang T, Shadlen MN (2007) Probabilistic reasoning by neurons. Nature 447:1075–1080

Yarkoni TBig correlations in little studies: inflated fMRI correlations reflect low statistical power-commentary on Vul, et al (2009) Perspect Psychol Sci 4:294–298

Zacks RT, Hasher L (2002) Frequency processing: a twenty-five year perspective. In: Sedlmeier PE, Betsch TE (eds) Frequency processing and cognition. Oxford University Press, New York, NY, pp 21–36

Zhang H, Maloney LT (2012) Ubiquitous log odds: a common representation of probability and frequency distortion in perception, action, and cognition. Front Neurosci 6:1

Printed in the United States
By Bookmasters